Sophia Haussener

Transport phenomena in complex multi-phase media

Sophia Haussener

Transport phenomena in complex multi-phase media

A tomography-based approach applied to solar fuel production and snow science

Südwestdeutscher Verlag für Hochschulschriften

Impressum / Imprint

Bibliografische Information der Deutschen Nationalbibliothek: Die Deutsche Nationalbibliothek verzeichnet diese Publikation in der Deutschen Nationalbibliografie; detaillierte bibliografische Daten sind im Internet über http://dnb.d-nb.de abrufbar.
Alle in diesem Buch genannten Marken und Produktnamen unterliegen warenzeichen-, marken- oder patentrechtlichem Schutz bzw. sind Warenzeichen oder eingetragene Warenzeichen der jeweiligen Inhaber. Die Wiedergabe von Marken, Produktnamen, Gebrauchsnamen, Handelsnamen, Warenbezeichnungen u.s.w. in diesem Werk berechtigt auch ohne besondere Kennzeichnung nicht zu der Annahme, dass solche Namen im Sinne der Warenzeichen- und Markenschutzgesetzgebung als frei zu betrachten wären und daher von jedermann benutzt werden dürften.

Bibliographic information published by the Deutsche Nationalbibliothek: The Deutsche Nationalbibliothek lists this publication in the Deutsche Nationalbibliografie; detailed bibliographic data are available in the Internet at http://dnb.d-nb.de.
Any brand names and product names mentioned in this book are subject to trademark, brand or patent protection and are trademarks or registered trademarks of their respective holders. The use of brand names, product names, common names, trade names, product descriptions etc. even without a particular marking in this works is in no way to be construed to mean that such names may be regarded as unrestricted in respect of trademark and brand protection legislation and could thus be used by anyone.

Coverbild / Cover image: www.ingimage.com

Verlag / Publisher:
Südwestdeutscher Verlag für Hochschulschriften
ist ein Imprint der / is a trademark of
OmniScriptum GmbH & Co. KG
Heinrich-Böcking-Str. 6-8, 66121 Saarbrücken, Deutschland / Germany
Email: info@svh-verlag.de

Herstellung: siehe letzte Seite /
Printed at: see last page
ISBN: 978-3-8381-3059-0

Zugl. / Approved by: Zürich, ETH, Diss., 2010

Copyright © 2012 OmniScriptum GmbH & Co. KG
Alle Rechte vorbehalten. / All rights reserved. Saarbrücken 2012

to my family

Acknowledgements

This book evolved from my doctoral thesis obtained at the Swiss Federal Institute of Technology Zurich (ETH). Therefore, I would like to thank Prof. Aldo Steinfeld for giving me the opportunity to do my thesis at the Professorship in Renewable Energy Carriers (PRE) and for supervising my thesis. The professional research environment he provided, the endless support, and his confidence in my work were essential to the success. I am very grateful to Prof. Jean Taine from the Laboratoire Énergétique, Moléculaire, Macroscopique, et Combustion of the École Central Paris for acting as co-examiner.

Various colleagues supported me in the experimental campaigns and the numerical tasks. I would like to thank Peter Wyss and Iwan Jerjen from the Swiss Federal Laboratories for Materials Science and Technology (EMPA) for granting me access to their tomographic facilities. Their expertise and support during the numerous computed tomography campaigns at the EMPA and at the Swiss Light Source (SLS) of the Paul Scherrer Institute (PSI) were highly valuable. I thank Alwin Frei for the support during my experimental gasification campaigns at the PSI. Martin Roeb and Dennis Thomey deserve special thanks for designing the solar evaporation/decomposition reactor and conducting the experimental campaign at the solar furnace of Deutsche Zentrum für Luft- und Raumfahrt (DLR) in Cologne. Their efficient work interaction made working with them a real pleasure for me. I would like to thank Martin Schneebeli for offering me the opportunity to do a research stay at the Swiss Federal Institute for Snow and Avalanche Research (SLF). The tomographic data of snow provided by the SLF is greatly appreciated. I thank Mathias Gergely of the SLF for the discussions on snow's radiation behavior and for sharing his experimental results on snow's transmission. I would like to thank the research team and Prof. Marco Stampanoni of the beamline for tomographic microscopy and coherent radiology experiments (TOMCAT) at the SLS for the support during the SLS campaigns. I thank Hansmartin Friess for developing the mesh generator tool and for continuously adapting it for my specific needs. Special thanks go to Jörg Petrasch, former PhD student at PRE, for starting the tomography-based research and providing me with the initial code. Wojciech Lipiński, former Postdoc at PRE, also deserves special thanks for the scientific discussions. In addition, I thank all visiting professors and researcher at PRE and the present and former students and members of PRE for providing a joyful research environment.

I would like to thank the Swiss Federal Institute of Technology Zurich for awarding the ETH Medal 2011 to my thesis. I would like to thank the Dimitris N. Chorafas Foundation for awarding the 2011 prize to my thesis.

The financial support by the Swiss National Science Foundation (contract no. 200021-115888) and the European Commission (contract no. 212470, project HycycleS) is gratefully acknowledged.

Finally, I would like to thank the persons close to my heart. Their unwavering support and confidence is the base of this work. I thank Daniela for showing me that everything is possible. I thank Flora, Julia and Gregor for giving me the courage for things I do.

Contents

1 Introduction	**1**
2 Volume averaging theory	**9**
2.1 Definitions	9
2.2 Heat transfer	11
2.2.1 Radiation	11
2.2.2 Conduction	21
2.2.3 Convection	22
2.2.4 Equation of state	22
2.3 Mass transfer	22
3 Methodology	**25**
3.1 Heat transfer	25
3.1.1 Radiation	25
3.1.2 Conduction	28
3.1.3 Convection	29
3.2 Mass transfer	30
3.2.1 Permeability and Dupuit-Forchheimer coefficient	30
3.2.2 Tortuosity and residence time distributions	30
3.2.3 Dispersion	31
4 Morphological characterization	**33**
4.1 Computed tomography	34
4.2 Segmentation and digitalization	35
4.3 Morphological properties	36
4.3.1 Numerical determination	36
4.3.2 Experimental determination	38
5 Reticulate porous ceramics	**41**
5.1 Sulfur-based water-splitting cycles	42
5.2 Characterization of reticulate porous ceramics	44

		5.2.1	Porous ceramics	44
		5.2.2	Morphological characterization	45
		5.2.3	Heat transfer characterization	48
		5.2.4	Mass transfer characterization	52
	5.3	Continuum model		54
		5.3.1	Model development	54
		5.3.2	Comparison to experiment	57
		5.3.3	Reference case	58
		5.3.4	Optimization	59
	5.4	Conclusions		63

6 Anisotropic ceramic foams — 65

- 6.1 Metal oxide/metal cycles with the non-stoichiometric CeO_2/Ce redox pair 66
- 6.2 Real samples 67
  - 6.2.1 Computed tomography 67
  - 6.2.2 Morphological characterization 68
  - 6.2.3 Heat transfer characterization 69
  - 6.2.4 Mass transfer characterization 72
- 6.3 Tailored foam design 74
- 6.4 Conclusions 76

7 Reacting packed bed of carbonaceous material — 77

- 7.1 Production of syngas by gasification of carbonaceous material . 78
- 7.2 Experimental campaign for sample production 82
- 7.3 Morphological characterization 83
  - 7.3.1 Computed tomography 83
  - 7.3.2 Porosity and specific surface area 85
  - 7.3.3 Particle-size distribution 87
  - 7.3.4 Representative elementary volume 88
- 7.4 Heat transfer characterization 88
  - 7.4.1 Radiative characterization 88
  - 7.4.2 Conductive characterization 91
  - 7.4.3 Convective characterization 91
- 7.5 Mass transfer characterization 95
  - 7.5.1 Permeability and Dupuit-Forchheimer coefficient 95
  - 7.5.2 Tortuosity and residence time 97
- 7.6 Conclusions 99

Contents

8 Semitransparent-particle packed bed   **101**
- 8.1 Thermal decomposition of calcium carbonate 102
- 8.2 Morphological characterization 103
  - 8.2.1 Computed tomography 103
  - 8.2.2 Porosity and specific surface 104
  - 8.2.3 Representative elementary volume 104
  - 8.2.4 Pore- and particle-size distributions 104
- 8.3 Radiative Properties 105
  - 8.3.1 Single-phase internal radiative properties 105
  - 8.3.2 Two-phase medium radiative coefficients 107
  - 8.3.3 Two-phase medium scattering phase functions 108
  - 8.3.4 Sensitivity analysis 109
  - 8.3.5 Accuracy and validation of the MC algorithm 110
- 8.4 Conclusions 112

9 Characterization of snow layers   **115**
- 9.1 Morphological characterization 116
  - 9.1.1 Computed tomography 116
  - 9.1.2 Porosity, specific surface area and REV 117
  - 9.1.3 Pore- and particle-size distributions 118
- 9.2 Radiative characterization 119
  - 9.2.1 Single-phase internal radiative properties 119
  - 9.2.2 Two-phase medium radiative properties 119
  - 9.2.3 Continuum properties 120
  - 9.2.4 Snow with soot impurities 125
- 9.3 Conclusions 128

10 Summary and outlook   **131**

Bibliography   **151**

Abstract

Transport phenomena in multi-phase media are of interest in a wide range of areas in science and industry chemical processing, combustion, nuclear and civil engineering, environmental and medical engineering, filtering and automotive applications, atmospheric sciences and solar engineering. Of special interest are solar thermal and thermochemical processes to generate electricity and (storable) solar fuels, solar materials and chemical commodities. In these processes multi-phase media serve as insulators, radiant absorbers, heat exchanges, catalyst carriers, reactant and reaction sites. The analysis of the complex interactions between multi-mode heat transfer, multiphase flow, and chemical reaction – on multiple scales – is fundamental to understanding and optimizing these processes. Volume averaging models for multi-phase media are commonly applied for process simulations. However, these models rely to a great extent on the accurate knowledge of the multi-phase media's effective transport properties, which in turn depend on the morphology and single phase properties.

A combined experimental-numerical procedure is adopted in this book: the exact 3D geometry of the complex multi-phase media are experimentally determined by computed tomography and used in direct discrete-scale simulations for morphological characterization and determination of the effective heat and mass transport properties. Two-point correlation functions and mathematical morphology operations are calculated for morphological characterization and validated by weight, BET surface area and laser scattering measurements. Collision-based Monte Carlo is utilized to calculate distribution functions of attenuation path length and direction of incidence at the phase boundary which are used to determine effective radiative properties. Spectroscopic measurements are conducted for validation of the calculated radiative properties. Finite volume techniques are used to solve for mass, momentum and energy conservation allowing for conductive/convective heat transfer and flow characterization. The methodology is then applied to four multi-phase media relevant in solar material and fuel processing: (i) reticulate porous ceramics, (ii) anisotropic ceramic foams, (*iii*) reacting packed beds and (*iv*) semitransparent-particle packed beds. A fifth medium relevant in the area of environmental sciences and climate modeling is investigated to show the wide applicability of the methodology: (*v*) layers of characteristic snow types.

Porosity, specific surface area, and pore-size distribution of a non-hollow SiSiC reticulate porous ceramic with nominal pore diameter of 1.27 mm are calculated. The determined extinction coefficient of 431 m^{-1} compares well to experimental estimates. The scattering phase function shows an enhanced fraction of backward scattering for assumed diffuse surface reflection. The ratio of effective conductivity to solid conductivity for small ratios of fluid to solid conductivities converges to 0.022. A Reynolds and Prandtl dependent Nusselt correlation is determined and converges to 6.8 for small Reynolds numbers. The numerically determined permeability and the Dupuit-Forchheimer coefficient compare well to values available in literature for materials with similar morphology. For neglected molecular dispersion, a Reynolds dependent dispersion tensor is calculated. Mean tortuosity is determined to be 1.07. The calculated effective properties are then incorporated in a continuum model of a solar evaporator/decomposer reactor. The reactor model is compared to experimentally measured temperatures at multiple locations within the reactor. Optimization of process parameters, reactor design and foam morphology is conducted; it shows a predominant influence of total acid solution flow rate and solar irradiation. Peak energetic and chemical efficiencies of 73% and 45%, respectively, are calculated for 2 ml/min acid inflow and 150 W solar power input.

Ceramic foams of ceria with structural anisotropy due to uniaxial pressing (along the z-direction) and anisotropic primary particles show enhanced extinction in z-direction because the pores are squeezed, which results in shorter attenuation paths. Effective conductivities in the x- and y-directions increase due to the more parallel alignment of the structures with the heat flux in these directions. Convective heat transfer in z-direction is larger because of the more tortuous path for fluid flow, increasing the accessible surface area for fluid-solid heat exchange. Reduced permeability and larger Dupuit-Forchheimer coefficient in z-direction are observed because of larger tortuosity along this direction. Preliminary studies on tailored foam designs, adjusted to the specific needs of the process in which the foam is applied, allow for foam engineering and consequently enhanced process performance.

A reacting packed bed is analyzed for heat and mass transfer. Gasification of waste tire shreds is chosen as model reaction. Porosity, specific surface and particle-size distribution are numerically calculated and compare to experimental data. Limitations in computerized tomography resolution shows to be the preliminary cause of discrepancies observed, especially at larger reaction extent where nanopores are formed. Larger extinction coefficients are calculated with increasing reaction extent due to particle shrinking and break-up, resulting in shorter attenuation path lengths. The scattering phase function shows to be independent of reaction extent for assumed diffusely reflecting particles. Effective conductivity, calculated by neglecting particle-particle contact resistances, decreases with reaction extent due to larger porosity and smaller particle size. The evolution of highly porous particles during pyrolysis and the subsequent shrinking and break-up of the particles cause a decrease and re-increase in convective heat exchange and Dupuit-Forchheimer coefficient. The largest permeability is calculated for the highly porous particles and decreases again for the final packed bed configuration. Comparison with heat and mass transfer properties available in the literature shows acceptable agreement for media with comparable morphology.

The derivation of the volume-averaged radiative transfer equations in multiphase media builds the basis for the radiative characterization of a semitransparent-particle packed bed. Nonspherical calcium carbonate particles are chosen as model particles and properties in the spectral range of 0.1 to 100 µm are calculated. The extinction coefficient for the transparent void phase solely depends on morphology, while the extinction coefficient for the semitransparent particles increases with increasing wavelength. The scattering coefficients in each phase are strongly influenced by the surface reflectivity of the boundary and show complementary behavior. The spectral scattering phase functions for diffusely reflecting particles show minor dependence on wavelength for both phases while they strongly depend on wavelength for specularly reflecting particles. Validation of the methodology by the analytical solutions for a diluted particle cloud of large opaque particles shows good agreement.

The morphological and radiative properties of snow layers composed of five characteristic snow types are numerically determined. Calculated extinction coefficients, scattering coefficients and scattering phase functions in the spectral range of 0.3 to 3 µm are then incorporated in a continuum model of a layer of snow composed of the different snow types and irradiated by a diffuse or collimated radiation flux. Overall reflectance, transmittance and absorptance are determined and compared to transmittance measured with a spectroscopic setup. Comparison of the calculated radiative properties based on the exact snow morphology, obtained by computerized tomography, with the one calculated based on simplified morphologies (packed beds of spheres) shows deviations up to 24% in reflectance, implying a significant influence of snow morphology on the radiative behavior. Additionally, soot impurities in snow are modeled and show a reduction in the calculated reflectivity by up to 83%.

The determined morphological properties can be used for the determination of structural parameters needed in kinetic models. The calculated effective heat and mass transport properties can be incorporated in volume-averaged (continuum) models of processes accounting for coupled heat/mass transfer (including chemical reactions) and fluid flow. The continuum models, in turn, are used for process design, modeling, optimization and scale-up. Accurate modeling and an in-depth understanding of the processes involving the multi-phase media is achieved. Additionally, the influence of multi-phase media's morphology on heat and mass transfer and consequently process performance is understood. The tomography-based discrete-scale numerical simulations show to be widely applicable, also for nonsolar applications such as environmental science and medial engineering.

Zusammenfassung

Transportphänomene in Mehrphasenmedien sind von grossem Interesse in Wissenschaft und Industrie, beispielweise in Gebieten der chemischen Verarbeitung, Verbrennung, im Nuklear- und Bauingenieurwesen, in Umweltnaturwissenschaften und Medizintechnik, für Anwendungen als Filter und in der Automobilindustrie, für Atmosphären- und Solarforschung. Von speziellem Interesse sind dabei solar betriebene thermische und thermochemische Prozesse zur Erzeugung von Elektrizität und (lagerbaren) solaren Brennstoffen, solaren Materialien und chemischen Rohstoffen. In jenen Prozessen dienen die Mehrphasenmedien als Isolatoren, Strahlungsabsorber, Wärmetauscher, Katalysatorträger, Reaktanten und Reaktionsort. Die Analyse der komplexen Interaktionen zwischen mehrartigem Wärmetransport, mehrphasiger Strömung und chemischer Reaktion – alles auf mehreren Grössenskalen – ist fundamental für das Verständnis und die Optimierung der Prozesse. Diese Prozesse werden mehrheitlich mit volumengemittelten Modellen simuliert. Diese Modelle beruhen jedoch auf der genauen Bestimmung der effektiven Transporteigenschaften der Mehrphasenmedien. Diese Transporteigenschaften wiederum hängen von der Morphologie des Mehrphasenmediums und den Eigenschaften der einzelnen Komponenten ab.

Im vorliegenden Buch wird eine Methode benützt, die numerische und experimentelle Verfahren kombiniert: Die exakte 3-D-Geometrie der komplexen Mehrphasenmedien wird mittels Computertomografie erlangt. Die exakte Geometrie wird dann in Simulationen auf den Grössenskalen der Poren angewendet und für die Berechnung der Morphologie sowie der effektiven Wärme- und Stofftransporteigenschaften benützt. Zweipunktekorrelationsfunktionen und mathematische Morphologie-Operationen werden berechnet um die Morphologie zu charakterisieren. Die experimentelle Validierung der Berechnungen erfolgt über Gewichtsmessung, Bestimmung der BET-Oberfläche und Laserstreuungsmessungen. Die kollisionsbasierte Monte-Carlo-Methode wird verwendet, um die Verteilungsfunktionen der Pfadlängen für Abschwächung und der Einfallsrichtungen auf der Phasengrenze zu berechnen. Jene Verteilungsfunktionen werden für die Bestimmung der effektiven Strahlungseigenschaften verwendet und diese mittels spektroskopischen Messungen validiert. Das Finite-Volumen-Verfahren wird für die Lösung der Massen-, Momenten- und Energieerhaltungsgleichungen benützt und ermöglicht die Bestimmung der Wärmeleitungs- und Konvektionseigenschaften. Auf gleiche Weise wird die Strömung charakterisiert. Die beschriebene Methode wird dann auf vier Medien, welche in solarer Brennstoff- und Materialherstellung relevant sind, angewandt: (i) netzartige poröse Keramik, (ii) anisotrope poröse Schäume, (iii) reagierende Schüttschichten und (iv) Schüttschichten aus semitransparenten Partikeln. Ein fünftes Medium, relevant im Bereich der Umweltnaturwissenschaften und Klimamodellierung, wird untersucht und zeigt die breite Anwendbarkeit der Methode: (v) Schichten aus charakteristischen Schneetypen.

Die Porösität, die spezifische Oberfläche und die Porengrössenverteilung von einer gefüllten netzartigen porösen Keramik aus SiSiC mit einem nominalen Porendurchmesser von 1.27 mm werden berechnet. Der berechnete Extinktionskoeffizient ist 431 m^{-1} und steht in gutem Vergleich zu den experimentellen Messungen. Die Phasenfunktion für eine diffus reflektierende Oberfläche weist einen erhöhten Anteil von Rückwärtsstreuung auf. Der Quotient von effektiver Wärmeleitung zur Wärmeleitung der festen Phase konvergiert zu dem Wert 0.022 für kleine Verhältnisse von Wärmeleitungskoeffizienten der flüssigen zur festen Phase. Eine Reynolds- und Prandtl-Zahl-abhängige Nusselt-Korrelation wird berechnet und konvergiert zu 6.8 für kleine Reynolds-Zahlen. Die numerisch berechnete Permeabilität und der numerisch berechnete Dupuit-Forchheimer-Koeffizient sind vergleichbar mit Werten aus der Literatur für Medien mit ähnlicher Morphologie. Ein Dispersionstensor, abhängig von der Reynolds-Zahl, wird berechnet, wobei die molekulare Dispersion vernachlässigt wird. Die gemittelte Tortuosität ist 1.07. Die berechneten effektiven Eigenschaften werden in einem Kontinuummodell des solaren Reaktors zur Verdampfung und Spaltung verwendet. Das Reaktormodell wird mit Temperaturmessungen an verschiedenen Orten im Reaktor verglichen. Die Optimierung der Prozessparameter, des Reaktordesigns und der Schaummorphologie wird durchgeführt und lässt den Schluss zu, dass die totale Säurelösungsflussrate und die solare Einstrahlung die Parameter mit dem grössten Einfluss sind. Spitzenwerte der energetischen und chemischen Effizienzen von 73% und 45% bei 2 ml/min Säurezufluss und 150 W solarer Einstrahlleistung werden berechnet.

Der Extinktionskoeffizient von keramischem Schaum aus Ceroxid, welches durch einachsiges Pressen (entlang der z-Richtung) und der Anisotropie der Primärpartikel strukturelle Anisotropie aufweist, ist entlang der z-Richtung erhöht. Dies weil Poren in dieser Richtung zusammengequetscht werden, was

Zusammenfassung

zu kürzeren Abschwächungslängen führt. Die effektiven Wärmeleitungen entlang der x- und y-Richtungen erhöhen sich, weil die Struktur sich parallel zu der Wärmeflussrichtung ausgerichtet hat. Der konvektive Wärmeübergang entlang der z-Richtung ist grösser, weil die Strömung grössere Windung erfährt und daher mehr Fläche für Wärmeaustausch zwischen Flüssigkeit und Feststoff zur Verfügung hat. Eine reduzierte Permeabilität und einen grösseren Dupuit-Forchheimer-Koeffiezienten entlang der z-Richtung werden wegen der erhöhten Tortuosität entlang dieser Richtung beobachtet. Eine einleitenden Studie über massgeschneiderte Schaumdesigns, welche den spezifischen Bedürfnissen der Prozesse, in welchen der Schaum verwendet wird, angepasst werden, ermöglicht es, Schaumanpassungen zu machen, die konsequenterweise die Prozesseffizienz erhöhen.

Die effektiven Wärme- und Stoffübertragungseigenschaften einer reagierenden Schüttschicht werden analysiert. Die Vergasung von zerkleinertem Abfallpneu wird als Modellreaktion gewählt. Die Porosität, die spezifische Oberfläche und die Partikelgrössenverteilung werden numerisch berechnet und mit experimentellen Daten verglichen. Limitierte Auflösung der Computertomografie ist der Hauptgrund für die Unterschiede zwischen den Berechnungen und den Experimenten, welche vor allem für höhere Reaktionskonversionen, bei welchen Nanoporen entstehen, beobachtet werden. Der Extinktionskoeffizient erhöht sich mit der Reaktionskonversion, da die Partikel schrumpfen und auseinanderbrechen. Dies resultiert in verkürzten Abschwächungslängen. Die Phasenfunktion verhält sich unabhängig von der Reaktionskonversion für diffus reflektierende Partikel. Die effektive Wärmeleitung, welche für vernachlässigte Partikel-Partikel-Kontaktwiderstände berechnet wird, verkleinert sich mit erhöhter Reaktionskonversion, da die Porosität zunimmt und die Partikel kleiner werden. Die Abnahme und die Wiederzunahme vom konvektiven Wärmeübergang und dem Dupuit-Forchheimer-Koeffizienten ist bedingt durch die Entstehung von hoch porösen Partikeln während der Pyrolyse und dem darauffolgenden Schrumpfen der Partikel und deren Auseinanderbrechen. Die grösste Permeabilität wird für die hoch porösen Partikel berechnet und nimmt wieder ab für die Konfiguration in der Schüttschicht am Ende der Reaktion. Vergleiche mit Wärme- und Stoffübertragungseigenschaften aus der Literatur zeigen akzeptable Übereinstimmung für ähnliche Morphologien.

Die Herleitung der volumengemittelten Strahlungsgleichungen in Mehrphasenmedien bilden die Basis für die Strahlungscharakterisierung einer Schüttschicht aus semitransparenten Partikeln. Nichtrunde Partikel aus Kalziumkarbonat werden als Modellpartikel gewählt. Die Eigenschaften werden in einem spektralen Bereich von 0.1 bis 100 µm berechnet. Der Extinktionskoeffizient für die transparente Phase hängt nur von der Morphologie ab. Der Extinktionskoeffizient für die feste Phase nimmt mit steigender Wellenlänge zu. Die Streuungskoeffizienten in jeder Phase sind stark abhängig vom Reflektionsverhalten der Phasengrenze und weisen komplementäres Verhalten auf. Die spektrale Phasenfunktion für diffus reflektierende Partikel zeigt für beide Phasen minimale Abhängigkeit von der Wellenlänge. Für spekular reflektierende Partikel sind sie stark wellenlängenabhängig. Validierung durch analytische Lösungen für verdünnte Partikelwolken, bestehend aus grossen undurchsichtigen Partikeln, weisen gute Übereinstimmung auf.

Die morphologischen und die Strahlungseigenschaften von Schneeschichten, bestehend je aus fünf charakteristischen Schneetypen, werden numerisch bestimmt. Die in einem spektralen Bereich von 0.3 bis 3 µm berechneten Extinktionskoeffizienten, Streuungskoeffizienten und Phasenfunktionen werden in einem Kontinuummodell einer Schneeschicht, welche aus verschiedenen Schneetypen besteht und mit diffuser oder kollimierter Strahlung beschienen wird, verwendet. Reflektivität, Transmission und Absorption der Strahlung werden bestimmt und mit Transmissionswerten, gemessen mit einer spektroskopischen Einrichtung, verglichen. Die Strahlungseigenschaften, welche basierend auf der exakten Morphologie, die mit Computertomografie ermittelt wird, berechnet werden, werden mit Strahlungseigenschaften verglichen, die basierend auf verein-fachter Morphologie (Schüttschicht von Kugeln) berechnet werden. Sie unterscheiden sich um bis zu 25% und zeigen auf, dass es einen signifikanten Einfluss der Schneemorphologie auf die Strahlungseigenschaften gibt. Zusätzlich werden Verunreinigungen aus Russ im Schnee modelliert und es wird gezeigt, dass diese Verunreinigungen die Reflektivität um bis zu 83% reduzieren können.

Die berechneten morphologischen Eigenschaften können für die Bestimmung von Strukturparametern verwendet werden, die in kinetischen Modellen benötigt werden. Die berechneten effektiven Wärme- und Stofftransporteigenschaften können in volumengemittelten Prozessmodellen verwendet werden, in welchen sie gekoppelten Wärme- und Stofftransport (mit chemischen Reaktionen) sowie Strömung berücksichtigen. Diese Kontinuummodelle werden dazu verwendet, die Prozesse zu dimensionieren, zu modellieren, zu optimieren und auf grössere Skalen auszulegen. Exakte Modellierung und ein detailliertes Verständnis der Prozesse, welche die Mehrphasenmedien beinhalten, wird im Verlaufe dieser Arbeit erzielt. Zusätzlich wird der Einfluss der Morphologie des Mehrphasenmediums auf die Wärme- und Stoffübertragungseigenschaften verstanden. Die tomografiebasierten Simulationen auf den Grössenskalen der Poren weisen eine breite Anwendbarkeit auf. Dies ist der Fall sowohl für solare als auch für nichtsolare Anwendungen wie beispielsweise im Bereich der Umweltnaturwissenschaften und der Medizintechnik.

Nomenclature

Symbols

A	surface area, m^2
	overall absorptance
A_0	specific surface, m^{-1}
a_i	fitting constants
b	bitnumber of image
c	constant, indicating gamma correction regime change
	molar concentration, molm^{-3}
c_0	speed of light in vacuum, 2.9979·10^8, ms^{-1}
c_p	specific heat capacity at constant pressure, Jkg^{-1}K^{-1}
D	dispersion tensor, m^{2}s^{-1}
d	particle diameter, m^{-1}
d_h	hydraulic diameter, m^{-1}
egf	electrical gain factor
em	specific emissions, kgkWh^{-1}
eo	specific electric output, kWhkg^{-1}
F	probability density function
F_DF	Dupuit-Forchheimer coefficient, m^{-1}
f	size distribution function, m^{-1}
	time distribution function, s^{-1}
f_v	volume fraction
f_γ	gamma correction function
G	cumulative distribution function
h	Planck's constant, 6.6261·10^{-34}, Js
	heat transfer coefficient, Wm^{-2}K^{-1}
	enthalpy, Jkg^{-1}
I	volume averaged radiative intensity, Wm^{-3}sr^{-1}
i, j	component indices

K	permeability, m²
k	imaginary part of complex refractive index
	conductivity, Wm⁻¹K⁻¹
k_B	Boltzmann's constant, $1.3807 \cdot 10^{-23}$, JK⁻¹
L	discrete-scale radiative intensity, Wm⁻³sr⁻¹
l	length, m
M	number of components
	molar mass, kgmol⁻¹
M_1	number of semitransparent components
m	complex refractive index
	mass, kg
$\dot{m}$	mass flow rate, kgs⁻¹
N	number of rays
	number of particles
N_i	number of components adjacted to component i
$N_{i,1}$	number of semitransparent components adjacted to component i
Nu	Nusselt number, $h_{sf}d_i k_f^{-1}$
n	real part of complex refractive index
	number of moles
	particle number density, m⁻³
$\dot{n}$	molar flow rate, mols⁻¹
p	(partial) pressure, Nm⁻²
Pr	Prandtl number, $c_p \mu k_f^{-1}$
q	heat rate, W
Q	efficiency factor
R	molar gas constant, 8.31447, Jmol⁻¹K⁻¹
	overall reflectance
Re	Reynolds number, $\langle u_i \rangle d_i \nu^{-1}$
$\Re$	random number between 0 and 1
r	distance between two points in the sample, m
r_{dif}	relative difference
$\mathbf{r}$	position vector for spatial coordinates in the sample
S_i	standard deviation of scalar function i
S_m	mass source, kgm⁻³s⁻¹
s	path length, m
s_2	two-point correlation function

$\hat{\mathbf{s}}$		unit vector of path direction
T		temperature, K
Tr		overall transmittance
t		time, s
$\mathbf{u}$		velocity vector, ms^{-1}
V		total sample volume, m^3
$\dot{V}$		volume flow rate, m^{3}s^{-1}
$\dot{W}$		work rate, W
X		reaction conversion
x		molar fraction
		mass fraction
		Cartesian coordinate
$\mathbf{x}$		spatial location vector
y		Cartesian coordinate
z		Cartesian coordinate

Greek symbols

α		absorption values of tomographic scans, m^{-1}
β		extinction coefficient, m^{-1}
γ_i		gamma correction factor
δ		half band with for REV calculations
ε		porosity
ε_r		emissivity
ϵ		normalized 2-norm of vector
η		efficiency
θ		angle, rad
κ		absorption coefficient, m^{-1}
λ		wavelength, m
μ		cosine of angle
		dynamic viscosity, kgm^{-1}s^{-1}
ν		kinematic viscosity, m^{2}s^{-1}
ξ		size parameter
		number fraction
Π_pg		dimensionless pressure gradient

ρ	density, kgm^{-3}	
ρ_r	reflectivity	
σ_s	scattering coefficient, m^{-1}	
τ	tortuosity	
	transmissivity	
Φ	scattering phase function	
φ_d	difference between azimutal angle of incidence and reflection, rad	
Ψ	pore scale indicator function	
ψ	scalar function	
Ω	solid angle, sr	

Subscripts

$\parallel$	parallel to heat flux or flow
$\perp$	perpendicular to heat flux or flow
a	absorption
agg	agglomerate
an	analytical
atm	atmospheric
att	attenuated
b	blackbody
C	carbon
cap	capillary
cond	conduction
conv	convection
d	discrete-scale
	diffuse
dep	dependent
decomp	decomposition
e	effective
	extinction
en	energy
eq	equivalent
evap	evaporation

ex	experimental
f	fluid phase
fib	fibrous
fs	feedstock
Gn	Gnielinski
Gu	Gunn
g	gas
ga	gasifying agent
hyd	hydraulic
i	incoming
imp	impurity
in	incident
	inflow
i, j, k	counters
K	Kozeny
l	liquid
lam	laminar
lm	logarithmic mean
λ	spectral
MD	MacDonald
m	mean
max	maximum
min	minimum
nom	nominal
num	numerical
op	opening
p	particle
R	Rumpf
r	radiation
refl	reflecting
Sa	Saidi
s	scattered
	solid phase
	sphere
sf	solid-fluid interface
sol	solar
sp	specular

sub	sub-micron
t	threshold
	transmitted
tot	total
turb	turbulent
vox	voxel
W	Ward
Wa	Wakao
w	continuum-scale medium boundary
	wall
wts	waste tire shreds
x, y, z	along x, y, or z, respectively
0	initial

Superscripts

$'$	fluctuation
$''$	bi-directional
	per area
$'''$	volumetric
$'\cap$	directional-hemispherical
$+$	forward direction
$-$	backward direction
f	fluid phase
i, j	counters
s	solid phase

Operators

$<.>$	superficial average
$<.>'$	intrinsic average
$\tilde{\ }$	fluctuations
$.'$	fluctuations

$\|\cdot\|_2$	two-norm
ln	natural logarithm
δ	Dirac delta
H	Heaviside step function
Δ	difference
∇	nabla symbol

Abbreviations

AG	autothermal gasification
BDOTS	bimodal distributed overlapping transparent sphere
BET	Brunauer-Emmett-Teller
CC	combined cycle
CPC	compound parabolic concentrator
CT	computed tomography
DLR	Deutsche Zentrum für Luft- und Raumfahrt
DO	discrete ordinate
DPLS	direct pore-level simulations
dh	depth hoar
ds	decomposed snow
EDS	energy dispersive X-ray spectrometer
EMPA	Swiss Federal Laboratories for Materials Science and Technology
ETH	Swiss Federal Institute of Technology
FC	fuel call
FOV	field of view
FV	finite volume
HR	high-resolution
ICSSG	International classification for seasonal snow on the ground
IOOS	identical overlapping opaque spheres
IOSS	identical overlapping semitransparent spheres
IOTS	identical overlapping transparent spheres
LHV	low heating value
IR	infrared
LR	low-resolution

MC	Monte Carlo	
m	metamorphosed	
ns	natural snow	
PRE	Professorship in Renewable Energy Carriers	
PSI	Paul Scherrer Institute	
ppi	pores per inch	
RC	Rankine cycle	
REV	representative elementary volume	
RMS	root mean square	
RPC	reticulate porous ceramic	
SG	steam gasification	
SLF	Swiss Federal Institute for Snow and Avalanche Research	
SLS	Swiss Light Source	
SOR	successive over-relaxation	
ss	snowmaker snow	
TOMCAT	tomographic microscopy and coherent radiology experiments	
WGS	water-gas shift reaction	
ws	wet snow	

Chapter 1

Introduction

The context of this book lies in the area of complex, chemically reacting multi-phase media in multi-scale systems used in solar energy applications. The achieved numerical characterization of the multi-phase media based on the exact morphology, obtained by computed tomography, provides an in-depth understanding of the processes' reaction mechanisms and of how morphology influences heat and mass transfer in multi-phase media. Additionally, it provides the basis for averaging models, which allow to investigate complex interactions of multi-mode heat transfer, multi-phase flow, and chemical reaction, and allow for subsequent optimization of the solar reactors and processes. Consequently, it is part of the effort from the transition of a fossil fuel based to a solar energy based society.

Two of the most pressing sustainability challenges in the 21st century, among others like clean water and food, are energy security and climate change. Today's primary energy supply and consumption are based on oil (33%), coal (27%) and gas (21%). They are ecologically, economically and socially unsustainable. The global primary energy demand is expected to increase further by 45% until 2030, driven by growth in world population and economy [3]. If the primary energy source continues to be mainly based on fossil energy carriers, the greenhouse gas emissions, in particular CO_2 emissions, are expected to increase by 45% and to cause an increase in global average temperature by approximately 2 °C. In the long term there is a 50% chance of global average temperature increase by more than 5 °C [155]. Serious changes in the ecosystem are expected and costs associated with climate change will increase exorbitantly [206]. Additionally, predicted decline in oil and gas production [3] further intensify the challenges and point to the need for alternative and sustainable primary energy sources.

Short term countermeasures to face the pressing challenges are increase in energy efficiency as well as carbon capturing and sequestration technologies. Long term strategies need to substitute the fossil energy sources by clean, renewable, and sustainable primary energy sources. Among these sources, namely biomass, wind, solar, water (including hydropower and ocean resources) and geothermal, the solar source exhibits the largest potential [114]. Covering 0.1% of the earth's land space (in a region with reasonable solar irradiation) with solar collectors operating with a collector efficiency of 20%, would provide enough energy to cover today's yearly global energy need. Since solar radiation is unequally distributed, intermittent and diluted solar radiation concentration and solar energy conversion and storage are crucial.

Concentration of solar radiation is achieved by optical systems such as through, fresnel, tower and dish systems [204, 77]. Solar radiation concentration ratios of few tens (through systems) up to several thousands (dish systems) are obtained. Solar energy is converted by using the concentrated solar radiation as source of process heat for thermal and thermochemical processes to produce electricity and (storable) solar fuels, solar materials, and chemical commodities.

A solar fuel of interest is hydrogen. Its solar production can be achieved by direct water or hydrogen sulfide splitting (a by-product from coal, petroleum and natural gas refining) [204, 202, 101]. The direct water-splitting requires temperatures above 2500 K, posing severe material problems and high re-radiation losses. Furthermore, it results in an explosive mixture of hydrogen and oxygen which need to be separated at high temperature to prevent recombination. Multi-step thermochemical water-splitting cycles are proposed to by-pass the separation problem and to allow an operation at relatively moderate temperatures. These cycles are often based on redox metallic systems. In order to reduce cycle inherent losses, associated to heat transfer and product separation at each step, two-step cycles are preferred. In general, an endothermic reduction step at high temperatures reduces the metal oxide, M_xO_y, to its metal or lower-valence metal oxide, M,

$$M_xO_y = xM + \frac{y}{2}O_2. \tag{1.1}$$

The process heat is provided by means of concentrated solar radiation. In the exothermic oxidation step, the metal or the lower-valence metal oxide is oxidized with steam to produce H_2 and the metal-oxide,

$$xM + yH_2O = M_xO_y + yH_2. \tag{1.2}$$

The latter is reused in the first step. Alternatively, CO_2 or a mixture of H_2O and CO_2 can be used as oxidant,

$$xM + yCO_2 = M_xO_y + yCO, \tag{1.3}$$

leading to the production of syngas, a mixture of mainly H_2 and CO with applications in power generation, fuels (gaseous and liquid), and basic chemical manufacturing. Using CO_2 captured from flue gases or atmospheric air [148] provides an opportunity to capture CO_2 and to produce carbon neutral solar fuels [123]. Three promising metallic redox pairs for two-step cycles have been investigated and experimentally demonstrated: Fe_3O_4/FeO [230, 28], ZnO/Zn [191, 159, 138], and CeO_2/Ce [1, 50].

Alternatively, three-step sulfur-iodine and two-step hybrid sulfur based cycles have been proposed [23, 22, 152, 154]. They need significant lower working temperatures. Initially, they have been proposed to be driven by waste heat from nuclear power plants. Recently, solar power has been investigated as the source of process heat [48, 181].

Since the replacement of fossil fuels and materials by solar fuels and materials requires the development of novel technologies and, consequently, longer to be economically competitive, mid-term goals have been formulated [203]. Hybrid solar/fossil processes, in which solar energy solely serves as source of process heat and the fossil energy exclusively serves as chemical source, are developed. Solar carbothermal reduction of metal oxides [124, 74, 205, 63] to produce metals and ammonia are exemplary processes for the generation of hybrid solar/fossil materials with lower energy intensity and CO_2 emission capacity compared to their conventional production. Fossil fuel free and CO_2 emission free solar materials can be produced by carbothermal metal oxide reduction processes if they are combined with syngas production processes [205, 75], and CO_2 neutral carbonaceous sources are used (such as charcoal, agricultural waste, or wood). Alternatively, at high temperatures direct metal oxide reduction is achieved [191, 123] without the need for chemical reducing agents eliminating CO_2 emissions.

Solar upgrading and decarbonization of fossil energy carriers by means of solar reforming, solar thermal cracking and solar gasification [41, 90, 241] are exemplary processes for the generation of hybrid solar/fossil fuels. Advantages of hybrid solar-/fossil-based processes compared to conventional (autothermal) cracking, reforming and gasification are fourfold: (i) the gaseous products are not contaminated by combustion's by-products; (ii) the discharge of pollutants to the environment is reduced; (iii) the calorific value of the feedstock is upgraded; and (iv) there is no need for energy-intensive processing of pure oxygen [217].

The solar reactors used for the production of solar materials and fuels often contain multi-phase[1] media such as packed beds [108, 166, 191], fluidized beds [224, 67, 148], particle-flows/aerosols [42, 242, 137, 117], monoliths [5, 150], and porous ceramics [53, 50]. The porous media serve as radiant absorbers, heat exchangers, catalyst carriers, reactant, reaction sites and insulators. Their heat and mass transfer properties are strongly influenced by, and dependent on, their morphology (porosity, connectivity of the porous network, size of the pores, dimension of the solid struts or particles, specific surface area, contact resistance etc.) and the discrete-scale physical properties. The effective properties are crucial for the reactor and, therefore, for the system performance.

Since multi-phase media are used not only in solar reactor technology but in a wide range of applications such as chemical processing, combustion, nuclear and civil engineering, environmental and medical engineering, filtering and automotive applications, and atmospheric sciences, the analysis of morphology, heat and mass transfer, and mechanical behavior in multi-phase media is of wide interest. Some examples are given: The mechanical properties of porous bone structures for fracture prevention [144], contaminant dispersion in soil for more effective ground water remediation strategies [188], permeability determination of anisotropic reservoir rock for enhanced hydrocarbon recovery [169], and the analysis of convective heat transfer in foams for development of high performance heat exchangers [19].

The characterization of multi-phase media and their optimization for specific applications is challenging because of their inherent structural complexity. Additionally, modeling systems including porous media, like solar reactors, deal with system-inherent scale disparity. The dimensions of the structure in the porous medium used in the solar reactor are in the micrometer scale, while the solar reactor has dimensions in the meter scale. It is impossible to resolve the smallest pores in a full reactor model. The same problem appears when modeling turbulent flows, where the scale of the smallest vortexes is much smaller than the system scale [171]. Therefore, as in turbulent flow theory, multi-scale approaches using averaging procedures (e.g., in time or space) have been introduced.

The volume-averaging method can be used to derive continuum equations for multi-phase systems. Essentially, conservation equations valid in each continuous phase are spatially smoothed to produce equations that are valid throughout the multi-phase media [195, 233, 102]. This can be done on multiple scales. The procedure is schematically shown in figure 1.1. The complexity of modeled sys-

---

[1]The expression multi-phase media is used interchangeably with porous media throughout this book.

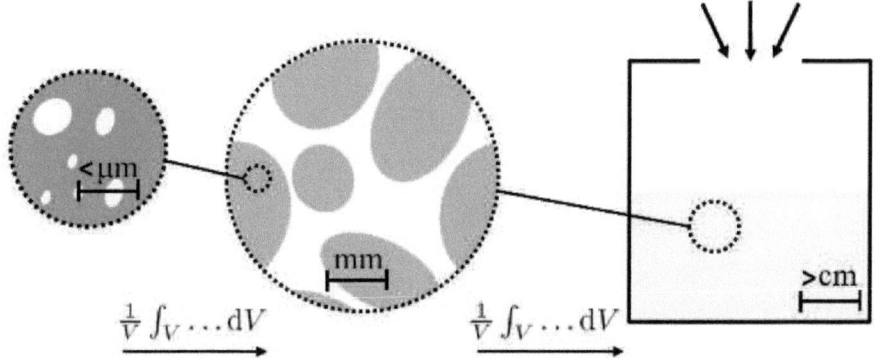

Figure 1.1: Principle of volume averaging on multi scales: a porous single particle being part of a packed bed, and the bed being part of a solar reactor.

tems, containing multi-phase media on multiple scales, is greatly reduced when using a volume averaging approach. Nevertheless, averaging poses a closure problem due to the occurrence of spatial deviation terms. These terms must be determined by solving additional equations. Often they are determined by introducing effective transport properties [233, 102]. These properties are strongly dependent on the phase boundaries and, consequently, on the discrete-scale geometry or morphology.

Non-destructive techniques, such as X-ray computed tomography (CT), can be used to obtain the exact discrete-scale geometry, which is then incorporated in direct discrete-scale numerical models for the determination of morphological, heat and mass transport, and mechanical properties [188].

In this book, a tomography-based approach for the characterization of multi-phase media [160] is extended in its applicability and used for the characterization, analysis and optimization of multi-phase media used in diverse solar thermal and thermochemical processes for the production of solar materials and fuels. A multi-phase medium relevant in the area of environmental science and climate modeling is additionally investigated to show the broad applicability of the methodology.

The thesis from which this book evolved is performed in the framework of the project HycycleS, funded by the European commission, and the SolRad project, funded by the Swiss National Science Foundation. The former involves collaborations with partners from the Deutsche Zentrum für Luft- und Raumfahrt (DLR, Germany), Commissariat à l'Energie Atomique (CEA, France), University of Sheffield (USFD, United Kingdom), Aerosol and Particle Technology

Laboratory (APTL, Greece), Joint Research Center Petten (JRC, Belgium), Ente per le Nuove tecnologie, l'Energia e l'Ambiente (ENEA, Italy), Empresarios Agrupados (EA, Spain), and Boostec Industries (Boostec, France). It is aimed at the qualification and enhancement of materials and components for key steps of sulfur-based thermochemical and thermochemical/electrochemical cycles for solar or nuclear hydrogen generation. Its final aim is to bring thermochemical water-splitting closer to realisation by improving the efficiency, stability, practicability, and costs of the key components involved and by elaborating engineering solutions. ETH's task is the detailed investigation of heat and mass transfer in the solar evaporator/decomposer reactor and subsequent optimization. The SolRad project on the other hand is aimed at the investigation of the fundamental heat and mass transfer phenomena in chemically reacting multi-phase media applied to the production of hydrogen via steam-gasification of carbonaceous materials under concentrated solar radiation.

In chapter 2, the basic definitions and fundamentals of the volume averaging theory are explained. The volume-averaged form of the conservation equations for multi-phase media and the definitions of the effective transport properties are given. The derivation of the volume-averaged radiative transfer equations (RTEs) are described in detail. Material from this chapter has been published in [120, 119, 161].

Chapter 3 describes the methodology used to determine the effective heat and mass transfer properties. The method based on radiative distribution function (RDF) identification is used for radiative characterization, calculated by collision-based Monte Carlo (MC) method on the discrete-scale. Finite volume (FV) technique is used for the determination of the effective heat conductivity by solving steady-state conduction equation on the discrete-scale. FV technique is used for calculating incompressible laminar flow and energy equation on the discrete-sale resulting in the determination of heat transfer coefficient, permeability, Dupuit-Frochheimer coefficient and dispersion tensor. The methodology to determine tortuosity and residence time distributions are additionally described, based on the flow field calculated on the discrete-scale.

The governing equations and the methodology for the calculation of the multi-phase media's morphological characterization, namely porosity, specific surface, pore- and particle-size distributions and representative elementary volume (REV), are detailed in chapter 4. Computed tomography, used for the experimental investigation of the exact morphology, is briefly introduced. Digital image processing and subsequent segmentation is then described, done to digitalize the exact morphology. Selected experimental methods for the validation of the numerical calculations of the morphological characteristics are explained.

Chapters 5 to 8 deal with the characterization of multi-phase media used in solar reactors for the production of solar materials and fuels. Two chapters are devoted to porous ceramic foams and two to packed beds. Chapter 9 deals with the characterization of snow, a multi-phase medium relevant in environmental science and climate modeling, and shows the broad applicability of the methodology in various other areas.

A sulfur-based thermochemical or hybrid thermo and electrochemical water-splitting cycle for the solar production of hydrogen is explained in chapter 5. The solar reactor developed for the high-temperature step incorporates reticulate porous ceramics used as radiant absorber, heat exchanger and reaction site. Morphological, heat and mass transfer characterization of the ceramic foam structure is achieved. These provide the input for the continuum model which is developed for subsequent reactor performance optimization. Experimental results from campaigns carried out at the solar furnace at DLR Cologne are used for comparison. Material from this chapter has been published in [212, 181, 78].

A ceria-based water-splitting cycle for the solar production of hydrogen is described in chapter 6. The solar reactor consists of a directly irradiated porous foam subjected to the reacting flow and therefore ensures efficient heat and mass transport to and from the surface reaction sites. They thereby maximize the kinetics of the reactions. Processing and primary particle structure of the ceria foam result in anisotropic micro-structure, which is characterized for morphology and heat and mass transfer properties. In a following first approach, artificially generated ceria samples are used for process optimization by adjusting the morphologies and, consequently, the transport properties to the specific process needs. Material from this chapter has been published in [84].

In chapter 7, the characterization of a reacting packed bed undergoing a solid-gas thermochemical transformation is presented. The production of syngas by gasification of carbonaceous waste material in a packed-bed reactor is chosen as model and describes a hybrid solar/fossil process for the production of a chemical commodity, or decarbonized and solar upgraded fuel. In contrast to the previously investigated ceramic foams the packed bed is a nonconsolidated structure and its morphology considerably changes during the reaction. The characterization of the packed bed and an in-depth understanding of the reaction mechanism is achieved. Material from this chapter has been published in [83, 81].

The production of lime by the solar-driven thermal decomposition of calcium carbonate as an example for the production of a solar material is considered in chapter 8. A conical rotary kiln, used for the solar-driven reaction, results in a packed-bed configuration of the semitransparent particles. In contrast to the previously investigated multi-phase media, radiation is able to partially pass through the particles in the packed bed. Morphology and radiative heat transfer properties are determined for the semitransparent packed bed. Material from this chapter has been published in [82].

Chapter 9 deals with an application in the area of environmental science and climate modeling: the spectral radiative characterization of snow, describing a packed bed of ice, water vapor and air. Discrete-scale radiative characteristics of five typical snow types is achieved. A continuum model of a slab of snow irradiated by diffuse or collimated radiation is developed. Results are validated by transmission measurement. The influence of grain shape and soot impurities on the radiation behavior is analyzed. Material from this chapter has been submitted for publication in [79].

Chapter 2
Volume averaging theory[1,2,3]

This chapter introduces the definitions of the volume averaging theory and briefly describes the averaging of the governing discrete-scale conservation equations, namely conservation of energy, mass, species and momentum, and subsequent closing of the derived volume-averaged equations. The averaging of the radiative transfer equation, representing conservation of intensity, is given in detail to allow for the derivation of the averaged divergence of the radiative flux term in the energy equation. The effective heat and mass transfer properties are introduced and can readily be used in the the volume-averaged continuum models.

2.1 Definitions

Consider the averaging volume V of a multi-component medium consisting of M components. Each component $i = 1, ..., M$ is characterized by its partial volume V_i and the corresponding volume fraction $f_{v,i}$[4]. A scalar quantity ψ_i for component i can be expressed as summation of its average and its fluctuation

$$\psi_i = \langle \psi_i \rangle + \psi_i', \tag{2.1}$$

---

[1]Material from this chapter has been published in: W. Lipiński, J. Petrasch, and S. Haussener. Application of the spatial averaging theorem to radiative heat transfer in two-phase media. *Journal of Quantitative Spectroscopy and Radiative Transfer*, 111:253–258, 2010. [120]

[2]Material from this chapter has been published in: W. Lipiński, D. Keene, S. Haussener, and J. Petrasch. Continuum radiative heat transfer modeling in media consisting of optically distinct components in the limit of geometrical optics. *Journal of Quantitative Spectroscopy and Radiative Transfer*, 111:2474–2480, 2010. [119]

[3]Material from this chapter has been published in: J. Petrasch, S. Haussener, and W. Lipiński. Discrete vs continuum level simulation of radiative transfer in semitransparent two-phase media. *Journal of Quantitative Spectroscopy and Radiative Transfer*, 112:1450–1459, 2011. [161]

[4]For void-solid two-phase media, $f_{v,\text{void}}$ is called porosity, ε.

where the superficial average, $\langle \psi_i \rangle$, is defined as

$$\langle \psi_i \rangle = \frac{1}{V} \int_V \psi_i \mathrm{d}V \, , V = \sum_{i=1}^{M} V_i \, . \qquad (2.2)$$

The intrinsic average, $\langle \psi_i \rangle^i$, is defined as

$$\langle \psi_i \rangle^i = \frac{1}{V_i} \int_{V_i} \psi_i \mathrm{d}V \, , V_i = f_{v,i} V \, , \qquad (2.3)$$

where

$$\psi_i = \langle \psi_i \rangle^i + \tilde{\psi}_i \, . \qquad (2.4)$$

A scalar quantity associated with a particular phase is required to vanish outside this phase. Thus,

$$\langle \psi_i \rangle = f_{v,i} \langle \psi_i \rangle^i \, . \qquad (2.5)$$

The relation between the superficial average of the discrete-scale gradient of ψ_i and the continuum-scale gradient of $\langle \psi_i \rangle$ is given by the spatial averaging theorem (SAT) [233, 102]:

$$\langle \nabla \psi_i \rangle = \nabla \langle \psi_i \rangle - \frac{1}{V} \sum_{j=1}^{N_i} \int_{A_{ij}} \psi_i \hat{\mathbf{n}}_{ji} \mathrm{d}A \, . \qquad (2.6)$$

A_{ij} is the interface surface area between component i and component j adjacent to i in the averaging volume V. N_i is the total number of components adjacent to the component i. $\hat{\mathbf{n}}_{ji}$ is the inner unit normal vector at the interface A_{ij}, i.e., $\hat{\mathbf{n}}_{ji}$ points into the component i. The transport theorem, which relates the average of a time derivative to the time derivative of the average, is given by

$$\left\langle \frac{\partial \psi_i}{\partial t} \right\rangle = \frac{\partial \langle \psi_i \rangle}{\partial t} + \frac{1}{V} \sum_{j=1}^{N_i} \int_{A_{ij}} \psi_i \mathbf{u}_{\mathrm{w},j} \cdot \hat{\mathbf{n}}_{ji} \mathrm{d}A \, . \qquad (2.7)$$

$\mathbf{u}_{\mathrm{w},j}$ is the velocity of the microscopic interface. For stationary, non-moving phase interfaces $\mathbf{u}_{\mathrm{w},j} = 0$.

Finally, the averaging volume V is assumed to be (i) sufficiently large to include all typical morphological structures of the multi-component medium and (ii) sufficiently small as compared to the overall size of the multi-component medium so that $\langle \psi_i \rangle$ and $\langle \nabla \psi_i \rangle$ can be assumed to be continuous scalar and vector fields, respectively.

2.2 Heat transfer

The energy equations of a two-phase medium, e.g., a connected moving fluid phase and a connected stationary semi-transparent rigid solid phase, are given for the fluid phase by

$$(\rho c_p)_\mathrm{f} \left[\frac{\partial T_\mathrm{f}}{\partial t} + \nabla \cdot (\mathbf{u}_\mathrm{f} T_\mathrm{f}) \right] = \nabla \cdot (k_\mathrm{f} \nabla T_\mathrm{f}) - \nabla \mathbf{q}''_{r,\mathrm{f}} + q'''_\mathrm{f}, \quad (2.8)$$

and the solid phase by

$$(\rho c_p)_\mathrm{s} \frac{\partial T_\mathrm{s}}{\partial t} = \nabla \cdot (k_\mathrm{s} \nabla T_\mathrm{s}) - \nabla \mathbf{q}''_{r,\mathrm{s}} + q'''_\mathrm{s}. \quad (2.9)$$

The terms q'''_i account for additional volumetric heat sinks or sources, e.g., by chemical reaction. At the phase boundary, temperatures and heat fluxes are assumed to be continuous. Volume averaging of equations (2.8) and (2.9) results in

$$(\rho c_p)_\mathrm{f} \left[\frac{\partial \langle T_\mathrm{f} \rangle}{\partial t} + \nabla \langle \mathbf{u}_\mathrm{f} T_\mathrm{f} \rangle \right] = \nabla \cdot (k_\mathrm{f} \nabla \langle T_\mathrm{f} \rangle) - \nabla \left[k_\mathrm{f} \frac{1}{V} \int_{A_\mathrm{sf}} T_\mathrm{f} \cdot \hat{\mathbf{n}}_\mathrm{sf} \mathrm{d}A \right]$$
$$- \frac{1}{V} \int_{A_\mathrm{sf}} k_\mathrm{f} \nabla T_\mathrm{f} \cdot \hat{\mathbf{n}}_\mathrm{sf} \mathrm{d}A - \nabla \langle \mathbf{q}''_{r,\mathrm{f}} \rangle + \frac{1}{V} \int_{A_\mathrm{sf}} \mathbf{q}''_{r,\mathrm{f}} \cdot \hat{\mathbf{n}}_\mathrm{sf} \mathrm{d}A + \langle q'''_\mathrm{f} \rangle, \quad (2.10)$$

$$(\rho c_p)_\mathrm{s} \frac{\partial \langle T_\mathrm{s} \rangle}{\partial t} = \nabla \cdot (k_\mathrm{s} \nabla \langle T_\mathrm{s} \rangle) + \nabla \left[k_\mathrm{s} \frac{1}{V} \int_{A_\mathrm{sf}} T_\mathrm{s} \cdot \hat{\mathbf{n}}_\mathrm{sf} \mathrm{d}A \right] + \frac{1}{V} \int_{A_\mathrm{sf}} k_\mathrm{s} \nabla T_\mathrm{s} \cdot \hat{\mathbf{n}}_\mathrm{sf} \mathrm{d}A$$
$$- \nabla \langle \mathbf{q}''_{r,\mathrm{s}} \rangle - \frac{1}{V} \int_{A_\mathrm{sf}} \mathbf{q}''_{r,\mathrm{s}} \cdot \hat{\mathbf{n}}_\mathrm{sf} \mathrm{d}A + \langle q'''_\mathrm{s} \rangle. \quad (2.11)$$

The first and second terms of the right hand side of eqs. (2.10) and (2.11) account for conduction. The third and fifth terms on the right hand side of eqs. (2.10) and (2.11) account for convection. The forth term on the right hand side of eqs. (2.10) and (2.11) describes the divergence of the averaged radiative heat flux vector across the medium. Since it involves the solution of additional conservation equations, namely conservation of intensity described by the radiative transfer equations (RTEs), it will be discussed in more detail.

2.2.1 Radiation

The continuum-scale divergence present in eqs. (2.10) and (2.11) is obtained by volume averaging of the discrete-scale RTEs in each phase[5] and subsequent integration over wavelength and solid angles of the obtained continuum-scale RTEs.

---
[5] The discrete-scale, superficial average and intrinsic average intensities associated with phase i are denoted by $L_i(\mathbf{r}, \hat{\mathbf{s}})$, $I_i(\mathbf{x}, \hat{\mathbf{s}}) = \langle L_i \rangle = \frac{1}{V} \int_V L_i \mathrm{d}V$, and $I_i^i(\mathbf{x}, \hat{\mathbf{s}}) = \langle L_i \rangle^i = \frac{1}{V_i} \int_{V_i} L_i \mathrm{d}V$, respectively.

Volume-averaged RTEs

A multi-component medium consisting of $i = 1, \ldots, M_1$ and $i = M_1 + 1, \ldots, M$ semi-transparent and opaque components, respectively, each of an arbitrary shape is considered. Each component i is adjacent to $j = 1, \ldots, N_{i,1}$ and $j = N_{i,1} + 1, \ldots, N_i$ semi-transparent and opaque components, respectively (see figure 2.1).

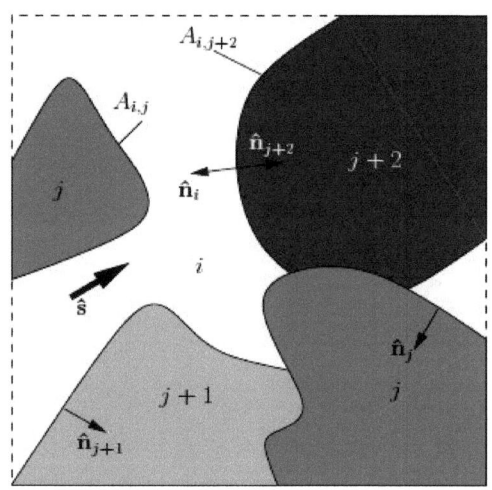

Figure 2.1: Multi-component medium with component designation.

The analysis presented in this section is subject to the following assumptions: (i) all components are isotropic; (ii) all components are non-polarizing and the state of polarization can be neglected; (iii) all components are at local thermodynamic equilibrium; (iv) characteristic dimensions of all components are much larger than the radiation wavelengths of interest so that laws of geometrical optics are valid in each component; (v) diffraction effects are negligible; (vi) dependent-scattering effects are negligible; (vii) all components are at rest as compared to the speed of light; ($viii$) radiative transfer in each component is quasi-steady.

Each component i is characterized by the set of the discrete-scale optical and radiative properties: the effective refractive index $n_i = $ const, the absorption and scattering coefficients, $\kappa_{d,i}$ and $\sigma_{s,d,i}$, respectively, and the scattering phase function $\Phi_{d,i}$. $\sigma_{s,d,i}$ and $\Phi_{d,i}$ are introduced to account for possible internal microscopic inhomogeneities of the components, and are to be determined by theories appropriate for characteristic length scales of the inhomogeneities.

2.2. Heat transfer

Furthermore, each component i is characterized by its temperature T_i, i.e. the components are allowed to be at thermal non-equilibrium with respect to each other.

The quasi-steady discrete-scale intensity in each component i can be determined by solving the corresponding quasi-steady discrete-scale RTEs [141]:

$$\hat{\mathbf{s}} \cdot \nabla L_i(\mathbf{r}, \hat{\mathbf{s}}) = -\beta_{\mathrm{d},i}(\mathbf{r}) L_i(\mathbf{r}, \hat{\mathbf{s}}) + \kappa_{\mathrm{d},i}(\mathbf{r}) L_{\mathrm{b},i}(\mathbf{r})$$
$$+ \frac{\sigma_{\mathrm{s,d},i}(\mathbf{r})}{4\pi} \int_{\Omega_\mathrm{i}=0}^{4\pi} L_i(\mathbf{r}, \hat{\mathbf{s}}_\mathrm{i}) \, \Phi_{\mathrm{d},i}(\mathbf{r}, \hat{\mathbf{s}}_\mathrm{i}, \hat{\mathbf{s}}) \, \mathrm{d}\Omega_\mathrm{i}, \quad i = 1, \ldots, M, \quad (2.12)$$

where the spectral subscript λ has been omitted for brevity. $L_{\mathrm{b},i}$ is the spectral blackbody intensity inside the component i. For the constant refractive index n_i, it is given by

$$L_{\mathrm{b},i}(\mathbf{r}) = \frac{2hc_0^2}{n_i^2 \lambda^5 \left[\exp\left(\frac{hc_0}{n_i k_{\mathrm{B}} \lambda T_i(\mathbf{r})} \right) - 1 \right]} \quad (2.13)$$

and λ is radiation wavelength in the component i. Equation (2.12) and, consequently, the following analysis require the effective refractive index to be constant within a component. Equation (2.12) is subject to the following boundary condition at $A_{ij, \hat{\mathbf{s}} \cdot \hat{\mathbf{n}}_{ji} > 0}$:

$$L_i(\mathbf{r}_{ij}, \hat{\mathbf{s}}) = \int_{\Omega_{\hat{\mathbf{s}}_\mathrm{i} \cdot \hat{\mathbf{n}}_{ji} > 0}} \tau_{ji}''(\mathbf{r}_{ij}, \hat{\mathbf{s}}_\mathrm{i}, \hat{\mathbf{s}}) L_j(\mathbf{r}_{ij}, \hat{\mathbf{s}}_\mathrm{i}) \hat{\mathbf{s}}_\mathrm{i} \cdot \hat{\mathbf{n}}_{ji} \, \mathrm{d}\Omega_\mathrm{i}$$
$$- \int_{\Omega_{\hat{\mathbf{s}}_\mathrm{i} \cdot \hat{\mathbf{n}}_{ji} < 0}} \rho_{\mathrm{r},ij}''(\mathbf{r}_{ij}, \hat{\mathbf{s}}_\mathrm{i}, \hat{\mathbf{s}}) L_i(\mathbf{r}_{ij}, \hat{\mathbf{s}}_\mathrm{i}) \hat{\mathbf{s}}_\mathrm{i} \cdot \hat{\mathbf{n}}_{ji} \, \mathrm{d}\Omega_\mathrm{i}, \quad j = 1, \ldots, N_i, \quad (2.14)$$

where $\mathbf{r}_{ij}$ is a position vector at the interface A_{ij}. The bi-directional reflection and transmission functions, $\rho_{\mathrm{r},ij}''$ and τ_{ij}'' respectively, must satisfy the condition of radiative intensity conservation for incidence at a reflecting-transmitting boundary:

$$\int_{\Omega_{\hat{\mathbf{s}} \cdot \hat{\mathbf{n}}_{ji} > 0}} \rho_{\mathrm{r},ij}''(\mathbf{r}_{ij}, \hat{\mathbf{s}}_\mathrm{i}, \hat{\mathbf{s}}) \hat{\mathbf{s}} \cdot \hat{\mathbf{n}}_{ji} \, \mathrm{d}\Omega - \int_{\Omega_{\hat{\mathbf{s}} \cdot \hat{\mathbf{n}}_{ji} < 0}} \tau_{ij}''(\mathbf{r}_{ij}, \hat{\mathbf{s}}_\mathrm{i}, \hat{\mathbf{s}}) \hat{\mathbf{s}} \cdot \hat{\mathbf{n}}_{ji} \, \mathrm{d}\Omega$$
$$= \rho_{\mathrm{r},ij}'^{\cap}(\mathbf{r}_{ij}, \hat{\mathbf{s}}_\mathrm{i}) + \tau_{ij}'^{\cap}(\mathbf{r}_{ij}, \hat{\mathbf{s}}_\mathrm{i}) = 1, \quad \hat{\mathbf{s}}_\mathrm{i} \cdot \hat{\mathbf{n}}_{ji} < 0. \quad (2.15)$$

For media with small discontinuities present between the components as compared to the radiation wavelengths of interest and to the characteristic dimensions of the components, $\rho_{\mathrm{r},ij}''$ and τ_{ij}'' are to be determined by including micro-scale radiative transfer effects at the interfaces. The intensity L_j in the boundary condition (2.14) can formally be obtained by solving eq. (2.12) for all components $j = 1, \ldots, N_i$. However, for opaque components $j = N_{i,1} + 1, \ldots, N_i$, the absorption coefficient approaches infinity, $\kappa_j \to \infty$, and the complete solutions to eq. (2.12) for these components are not required. A closer inspection of eq. (2.12) shows that the intensity $L_j(\mathbf{r}_{ij}, \hat{\mathbf{s}})$ results only from local emission within the component j in the vicinity of the interface. This allows rewriting eqs. (2.12) and (2.14) as

$$\hat{\mathbf{s}} \cdot \nabla L_i(\mathbf{r}, \hat{\mathbf{s}}) = -\beta_{\mathrm{d},i}(\mathbf{r}) L_i(\mathbf{r}, \hat{\mathbf{s}}) + \kappa_{\mathrm{d},i}(\mathbf{r}) L_{\mathrm{b},i}(\mathbf{r})$$
$$+ \frac{\sigma_{\mathrm{s,d},i}(\mathbf{r})}{4\pi} \int_{\Omega_{\mathrm{i}}=0}^{4\pi} L_i(\mathbf{r}, \hat{\mathbf{s}}_{\mathrm{i}}) \, \Phi_{\mathrm{d},i}(\mathbf{r}, \hat{\mathbf{s}}_{\mathrm{i}}, \hat{\mathbf{s}}) \, \mathrm{d}\Omega_{\mathrm{i}}, \qquad i = 1, \ldots, M_1, \qquad (2.16)$$

$$L_i(\mathbf{r}_{ij}, \hat{\mathbf{s}}) = \int_{\Omega_{\hat{\mathbf{s}}_{\mathrm{i}} \cdot \hat{\mathbf{n}}_{ji} > 0}} \tau''_{ji}(\mathbf{r}_{ij}, \hat{\mathbf{s}}_{\mathrm{i}}, \hat{\mathbf{s}}) L_j(\mathbf{r}_{ij}, \hat{\mathbf{s}}_{\mathrm{i}}) \, \hat{\mathbf{s}}_{\mathrm{i}} \cdot \hat{\mathbf{n}}_{ji} \, \mathrm{d}\Omega_{\mathrm{i}}$$
$$- \int_{\Omega_{\hat{\mathbf{s}}_{\mathrm{i}} \cdot \hat{\mathbf{n}}_{ji} < 0}} \rho''_{\mathrm{r},ij}(\mathbf{r}_{ij}, \hat{\mathbf{s}}_{\mathrm{i}}, \hat{\mathbf{s}}) L_i(\mathbf{r}_{ij}, \hat{\mathbf{s}}_{\mathrm{i}}) \, \hat{\mathbf{s}}_{\mathrm{i}} \cdot \hat{\mathbf{n}}_{ji} \, \mathrm{d}\Omega_{\mathrm{i}}, \qquad j = 1, \ldots, N_{i,1}, \qquad (2.17\mathrm{a})$$

$$L_i(\mathbf{r}_{ij}, \hat{\mathbf{s}}) = \varepsilon'_{\mathrm{r},ji}(\mathbf{r}_{ij}, \hat{\mathbf{s}}) L_{\mathrm{b},j}(\mathbf{r}_{ij}) - \int_{\Omega_{\hat{\mathbf{s}}_{\mathrm{i}} \cdot \hat{\mathbf{n}}_{ji} < 0}} \rho''_{\mathrm{r},ij}(\mathbf{r}_{ij}, \hat{\mathbf{s}}_{\mathrm{i}}, \hat{\mathbf{s}}) L_i(\mathbf{r}_{ij}, \hat{\mathbf{s}}_{\mathrm{i}}) \, \hat{\mathbf{s}}_{\mathrm{i}} \cdot \hat{\mathbf{n}}_{ji} \, \mathrm{d}\Omega_{\mathrm{i}},$$
$$j = N_{i,1} + 1, \ldots, N_i, \qquad (2.17\mathrm{b})$$

where $\varepsilon'_{\mathrm{r},ji}$ is the directional spectral emissivity of the interface between components j and i defined as

$$\varepsilon'_{\mathrm{r},ji}(\mathbf{r}_{ij}, \hat{\mathbf{s}}) \frac{L_{\mathrm{e},i}(\mathbf{r}_{ij}, \hat{\mathbf{s}})}{L_{\mathrm{b},j}(\mathbf{r}_{ij})}. \qquad (2.18)$$

$L_{\mathrm{b},j}(\mathbf{r}_{ij})$ is the blackbody intensity emitted by the component j into the component i,

$$L_{\mathrm{b},j}(\mathbf{r}_{ij}) = \frac{2hc_0^2}{n_i^2 \lambda^5 \left[\exp\left(\frac{hc_0}{n_i k_{\mathrm{B}} \lambda T_j(\mathbf{r}_{ij})}\right) - 1\right]}. \qquad (2.19)$$

Note that the interface emissivity $\varepsilon'_{\mathrm{r},ji}(\mathbf{r}_{ij}, \hat{\mathbf{s}})$ captures the interface transmission from the component j into the component i and the definition (2.18) follows the definition of the emissivity of opaque surfaces [141].

The condition of radiative intensity conservation at the interface between the semi-transparent component i and the opaque component j can be written

2.2. Heat transfer

analogously to eq. (2.15),

$$\int_{\Omega_{\hat{s}\cdot\hat{n}_{ji}>0}} \rho''_{\mathrm{r},ij}\left(\mathbf{r}_{ij},\hat{\mathbf{s}}_{\mathrm{i}},\hat{\mathbf{s}}\right)\hat{\mathbf{s}}\cdot\hat{\mathbf{n}}_{ji}\,\mathrm{d}\Omega + \alpha'_{ij}\left(\mathbf{r}_{ij},\hat{\mathbf{s}}_{\mathrm{i}}\right)$$

$$= \rho'^{\cap}_{\mathrm{r},ij}\left(\mathbf{r}_{ij},\hat{\mathbf{s}}_{\mathrm{i}}\right) + \alpha'_{ij}\left(\mathbf{r}_{ij},\hat{\mathbf{s}}_{\mathrm{i}}\right) = 1, \qquad \hat{\mathbf{s}}_{\mathrm{i}}\cdot\hat{\mathbf{n}}_{ji}<0, \qquad (2.20)$$

where Kirchhoff's law,

$$\alpha'_{ij}\left(\mathbf{r},\hat{\mathbf{s}}_{\mathrm{i}}\right) = \varepsilon'_{\mathrm{r},ji}\left(\mathbf{r},-\hat{\mathbf{s}}_{\mathrm{i}}\right), \qquad (2.21)$$

has been applied. Applying eq. (2.2) to each term of eq. (2.16) leads to:

$$\langle\hat{\mathbf{s}}\cdot\nabla L_i\left(\mathbf{r},\hat{\mathbf{s}}\right)\rangle = -\langle\beta_{\mathrm{d},i}\left(\mathbf{r}\right)L_i\left(\mathbf{r},\hat{\mathbf{s}}\right)\rangle + \langle\kappa_{\mathrm{d},i}\left(\mathbf{r}\right)L_{\mathrm{b},i}\left(\mathbf{r}\right)\rangle$$
$$+ \frac{1}{4\pi}\left\langle\sigma_{\mathrm{s,d},i}\left(\mathbf{r}\right)\int_{\Omega_{\mathrm{i}}=0}^{4\pi} L_i\left(\mathbf{r},\hat{\mathbf{s}}_{\mathrm{i}}\right)\Phi_{\mathrm{d},i}\left(\mathbf{r},\hat{\mathbf{s}}_{\mathrm{i}},\hat{\mathbf{s}}\right)\mathrm{d}\Omega_{\mathrm{i}}\right\rangle, \quad i=1,\ldots,M_1. \qquad (2.22)$$

The left-hand side of eq. (2.22) is developed by applying SAT, eq. (2.6). The terms on the right-hand side of eq. (2.22) are developed by assuming that the variation of the discrete-scale radiative properties is small enough so that the corresponding variation of the discrete scale radiative properties inside the averaging volume V becomes negligible [119]. Interchanging the order of integration with respect to the solid angle Ω_{i} and the volume V in the incoming scattering term results in:

$$\hat{\mathbf{s}}\cdot\nabla I_i\left(\mathbf{x},\hat{\mathbf{s}}\right) = -\beta_{\mathrm{d},i}\left(\mathbf{x}\right)I_i\left(\mathbf{x},\hat{\mathbf{s}}\right) + \kappa_{\mathrm{d},i}\left(\mathbf{x}\right)I_{\mathrm{b},i}\left(\mathbf{x}\right)$$
$$+ \frac{\sigma_{\mathrm{s,d},i}\left(\mathbf{x}\right)}{4\pi}\int_{\Omega_{\mathrm{i}}=0}^{4\pi} I_i\left(\mathbf{x},\hat{\mathbf{s}}_{\mathrm{i}}\right)\Phi_{\mathrm{d},i}\left(\mathbf{x},\hat{\mathbf{s}}_{\mathrm{i}},\hat{\mathbf{s}}\right)\mathrm{d}\Omega_{\mathrm{i}} + \frac{1}{V}\sum_{j=1}^{N_i}\int_{A_{ij}} L_i\left(\mathbf{r},\hat{\mathbf{s}}\right)\hat{\mathbf{s}}\cdot\hat{\mathbf{n}}_{ji}\,\mathrm{d}A,$$
$$i=1,\ldots,M_1. \qquad (2.23)$$

The last term on the right-hand side of eq. (2.23) represents the contribution to the spectral superficial average radiative heat transfer rate per unit volume and solid angle around the direction $\hat{\mathbf{s}}$ by radiative intensity in component i launched at $A_{ij,\hat{\mathbf{s}}\cdot\hat{\mathbf{n}}_{ji}>0}$ into the direction $\hat{\mathbf{s}}$ and removed in component i at $A_{ij,\hat{\mathbf{s}}\cdot\hat{\mathbf{n}}_{ji}<0}$ from the direction $\hat{\mathbf{s}}$, respectively. Thus, the surface integral in eq. (2.23) is split into two parts,

$$\int_{A_{ij}} L_i\left(\mathbf{r},\hat{\mathbf{s}}\right)\hat{\mathbf{s}}\cdot\hat{\mathbf{n}}_{ji}\,\mathrm{d}A = \int_{A_{ij,\hat{\mathbf{s}}\cdot\hat{\mathbf{n}}_{ji}>0}} L_i\left(\mathbf{r},\hat{\mathbf{s}}\right)\hat{\mathbf{s}}\cdot\hat{\mathbf{n}}_{ji}\,\mathrm{d}A + \int_{A_{ij,\hat{\mathbf{s}}\cdot\hat{\mathbf{n}}_{ji}<0}} L_i\left(\mathbf{r},\hat{\mathbf{s}}\right)\hat{\mathbf{s}}\cdot\hat{\mathbf{n}}_{ji}\,\mathrm{d}A. \qquad (2.24)$$

The first term on the right-hand side of eq. (2.24) is developed by using the boundary conditions, eqs. (2.17a) and (2.17b). The second term on the right-hand side of eq. (2.24) is developed by using eqs. (2.15) and (2.20). Equation (2.24) becomes:

$$\int_{A_{ij}} L_i(\mathbf{r},\hat{\mathbf{s}})\,\hat{\mathbf{s}}\cdot\hat{\mathbf{n}}_{ji}\,\mathrm{d}A = \sum_{j=N_{i,1}+1}^{N_i} \int_{A_{ij,\hat{\mathbf{s}}\cdot\hat{\mathbf{n}}_{ji}>0}} \varepsilon'_{\mathrm{r},ji}(\mathbf{r}_{ij},\hat{\mathbf{s}})\,L_{\mathrm{b},j}(\mathbf{r}_{ij})\,\hat{\mathbf{s}}\cdot\hat{\mathbf{n}}_{ji}\,\mathrm{d}A$$

$$-\sum_{j=1}^{N_i}\int_{A_{ij,\hat{\mathbf{s}}\cdot\hat{\mathbf{n}}_{ji}>0}}\int_{\Omega_{\hat{\mathbf{s}}_\mathrm{i}\cdot\hat{\mathbf{n}}_{ji}<0}} \rho''_{\mathrm{r},ij}(\mathbf{r},\hat{\mathbf{s}}_\mathrm{i},\hat{\mathbf{s}})\,L_i(\mathbf{r},\hat{\mathbf{s}}_\mathrm{i})\,\hat{\mathbf{s}}_\mathrm{i}\cdot\hat{\mathbf{n}}_{ji}\,\mathrm{d}\Omega_\mathrm{i}\hat{\mathbf{s}}\cdot\hat{\mathbf{n}}_{ji}\,\mathrm{d}A$$

$$+\sum_{j=1}^{N_{i,1}}\int_{A_{ij,\hat{\mathbf{s}}\cdot\hat{\mathbf{n}}_{ji}>0}}\int_{\Omega_{\hat{\mathbf{s}}_\mathrm{i}\cdot\hat{\mathbf{n}}_{ji}>0}} \tau''_{ji}(\mathbf{r},\hat{\mathbf{s}}_\mathrm{i},\hat{\mathbf{s}})\,L_j(\mathbf{r},\hat{\mathbf{s}}_\mathrm{i})\,\hat{\mathbf{s}}_\mathrm{i}\cdot\hat{\mathbf{n}}_{ji}\,\mathrm{d}\Omega_\mathrm{i}\hat{\mathbf{s}}\cdot\hat{\mathbf{n}}_{ji}\,\mathrm{d}A$$

$$+\sum_{j=N_{i,1}+1}^{N_i}\int_{A_{ij,\hat{\mathbf{s}}\cdot\hat{\mathbf{n}}_{ji}<0}} \alpha'_{ij}(\mathbf{r},\hat{\mathbf{s}})\,L_i(\mathbf{r},\hat{\mathbf{s}})\,\hat{\mathbf{s}}\cdot\hat{\mathbf{n}}_{ji}\,\mathrm{d}A$$

$$+\sum_{j=1}^{N_i}\int_{A_{ij,\hat{\mathbf{s}}\cdot\hat{\mathbf{n}}_{ji}<0}} \rho'^{\cap}_{\mathrm{r},ij}(\mathbf{r},\hat{\mathbf{s}})\,L_i(\mathbf{r},\hat{\mathbf{s}})\,\hat{\mathbf{s}}\cdot\hat{\mathbf{n}}_{ji}\,\mathrm{d}A$$

$$+\sum_{j=1}^{N_{i,1}}\int_{A_{ij,\hat{\mathbf{s}}\cdot\hat{\mathbf{n}}_{ji}<0}} \tau'^{\cap}_{ij}(\mathbf{r},\hat{\mathbf{s}})\,L_i(\mathbf{r},\hat{\mathbf{s}})\,\hat{\mathbf{s}}\cdot\hat{\mathbf{n}}_{ji}\,\mathrm{d}A\,. \tag{2.25}$$

The first three terms on the right-hand side of eq. (2.25) quantify the augmentation of the superficial average radiative heat transfer rate per unit solid angle around direction $\hat{\mathbf{s}}$ in the component i resulting from the intensities emitted, reflected, and transmitted into the component i, respectively. The last three terms on the right-hand side of eq. (2.25) quantify the attenuation of the superficial average radiative heat transfer rate per unit solid angle around direction $\hat{\mathbf{s}}$ by interface absorption, reflection, and transmission, respectively. The following absorption and scattering coefficients, and the scattering phase functions

2.2. Heat transfer

associated with the superficial average intensities I_i and $I_{\text{b},j}$ are postulated:

$$\kappa_{ij}\left(\mathbf{x}\right) = -\frac{\displaystyle\int_{A_{ij,\hat{\mathbf{s}}\cdot\hat{\mathbf{n}}_{ji}<0}} \alpha'_{ij}\left(\mathbf{r}_{ij},\hat{\mathbf{s}}\right) L_i\left(\mathbf{r},\hat{\mathbf{s}}\right) \hat{\mathbf{s}}\cdot\hat{\mathbf{n}}_{ji}\,\mathrm{d}A}{I_i\left(\mathbf{x},\hat{\mathbf{s}}\right) V}, \tag{2.26}$$

$$\kappa_{ji}\left(\mathbf{x}\right) = \frac{\displaystyle\int_{A_{ij,\hat{\mathbf{s}}\cdot\hat{\mathbf{n}}_{ji}>0}} \varepsilon'_{\text{r},ji}\left(\mathbf{r}_{ij},\hat{\mathbf{s}}\right) L_{\text{b},j}\left(\mathbf{r},\hat{\mathbf{s}}\right) \hat{\mathbf{s}}\cdot\hat{\mathbf{n}}_{ji}\,\mathrm{d}A}{I_{\text{b},j}\left(\mathbf{x},\hat{\mathbf{s}}\right) V}, \tag{2.27}$$

$$\sigma_{\text{s,refl},i}\left(\mathbf{x}\right) = -\frac{\displaystyle\int_{A_{ij,\hat{\mathbf{s}}\cdot\hat{\mathbf{n}}_{ji}<0}} \rho'^{\cap}_{\text{r},ij}\left(\mathbf{r},\hat{\mathbf{s}}\right) L_i\left(\mathbf{r},\hat{\mathbf{s}}\right) \hat{\mathbf{s}}\cdot\hat{\mathbf{n}}_{ji}\,\mathrm{d}A}{I_i\left(\mathbf{x},\hat{\mathbf{s}}\right) V}, \tag{2.28}$$

$$\sigma_{\text{s},ij}\left(\mathbf{x}\right) = -\frac{\displaystyle\int_{A_{ij,\hat{\mathbf{s}}\cdot\hat{\mathbf{n}}_{ji}<0}} \tau'^{\cap}_{ij}\left(\mathbf{r},\hat{\mathbf{s}}\right) L_i\left(\mathbf{r},\hat{\mathbf{s}}\right) \hat{\mathbf{s}}\cdot\hat{\mathbf{n}}_{ji}\,\mathrm{d}A}{I_i\left(\mathbf{x},\hat{\mathbf{s}}\right) V}, \tag{2.29}$$

$$\Phi_{\text{refl},i}\left(\mathbf{x},\hat{\mathbf{s}}_{\text{i}},\hat{\mathbf{s}}\right) = -\frac{\displaystyle\int_{A_{ij,\hat{\mathbf{s}}\cdot\hat{\mathbf{n}}_{ji}>0}} \rho''_{\text{r},ij}\left(\mathbf{r},\hat{\mathbf{s}}_{\text{i}},\hat{\mathbf{s}}\right) L_i\left(\mathbf{r},\hat{\mathbf{s}}_{\text{i}}\right) \hat{\mathbf{s}}_{\text{i}}\cdot\hat{\mathbf{n}}_{ji}\hat{\mathbf{s}}\cdot\hat{\mathbf{n}}_{ji}\,\mathrm{d}A}{\left(4\pi\right)^{-1}\sigma_{\text{s,r},ij}I_i\left(\mathbf{x},\hat{\mathbf{s}}\right) V}, \hat{\mathbf{s}}_{\text{i}}\cdot\hat{\mathbf{n}}_{ji} < 0, \tag{2.30}$$

$$\Phi_{ji}\left(\mathbf{x},\hat{\mathbf{s}}_{\text{i}},\hat{\mathbf{s}}\right) = \frac{\displaystyle\int_{A_{ij,\hat{\mathbf{s}}\cdot\hat{\mathbf{n}}_{ji}>0}} \tau''_{ji}\left(\mathbf{r},\hat{\mathbf{s}}_{\text{i}},\hat{\mathbf{s}}\right) L_j\left(\mathbf{r},\hat{\mathbf{s}}_{\text{i}}\right) \hat{\mathbf{s}}_{\text{i}}\cdot\hat{\mathbf{n}}_{ji}\hat{\mathbf{s}}\cdot\hat{\mathbf{n}}_{ji}\,\mathrm{d}A}{\left(4\pi\right)^{-1}\sigma_{\text{s,t},ji}I_j\left(\mathbf{x},\hat{\mathbf{s}}\right) V}, \hat{\mathbf{s}}_{\text{i}}\cdot\hat{\mathbf{n}}_{ji} > 0. \tag{2.31}$$

Note that the blackbody intensity $I_{\text{b},j}$ appearing in eq. (2.27) is obtained by applying eq. (2.2) to a discrete-scale blackbody intensity that would fill the component i as a result of emission from a black interface $A_{ij,\hat{\mathbf{s}}\cdot\hat{\mathbf{n}}_{ji}>0}$ into the component i, $\varepsilon'_{\text{r},ji} = 1$. Substituting eqs. (2.24) to (2.31) into eq. (2.23) and omitting for brevity the position vector notation in the radiative properties results in:

$$\hat{\mathbf{s}} \cdot \nabla I_i(\mathbf{x}, \hat{\mathbf{s}}) = -\left[\beta_{\mathrm{d},i} + \sum_{j=N_{i,1}+1}^{N_i} \kappa_{ij} + \sum_{j=1}^{N_{i,1}} \sigma_{\mathrm{s},ij} + \sum_{j=1}^{N_i} \sigma_{\mathrm{s,refl},i}\right] I_i(\mathbf{x}, \hat{\mathbf{s}}) + \kappa_{\mathrm{d},i} I_{\mathrm{b},i}(\mathbf{x}, \hat{\mathbf{s}})$$

$$+ \sum_{j=N_{i,1}+1}^{N_i} \kappa_{ji} I_{\mathrm{b},j}(\mathbf{x}, \hat{\mathbf{s}}) + \frac{\sigma_{\mathrm{s,d},i}}{4\pi} \int_{\Omega_i=0}^{4\pi} I_i(\mathbf{x}, \hat{\mathbf{s}}_i) \Phi_{\mathrm{d},i}(\hat{\mathbf{s}}_i, \hat{\mathbf{s}}) \, \mathrm{d}\Omega_i$$

$$+ \sum_{j=1}^{N_{i,1}} \frac{\sigma_{\mathrm{s},ji}}{4\pi} \int_{\Omega_i=0}^{4\pi} I_j(\mathbf{x}, \hat{\mathbf{s}}_i) \Phi_{ji}(\hat{\mathbf{s}}_i, \hat{\mathbf{s}}) \, \mathrm{d}\Omega_i$$

$$+ \sum_{j=1}^{N_i} \frac{\sigma_{\mathrm{s,refl},i}}{4\pi} \int_{\Omega_i=0}^{4\pi} I_i(\mathbf{x}, \hat{\mathbf{s}}_i) \Phi_{\mathrm{refl},i}(\hat{\mathbf{s}}_i, \hat{\mathbf{s}}) \, \mathrm{d}\Omega_i, \qquad i=1,\ldots,M_1. \qquad (2.32)$$

The set of eqs. (2.32) presents generalization of eq. (1) in [238], eqs. (8) and (9) in [72], and eqs. (18) and (27) in [120], for a multi-component medium consisting of any number of semi-transparent and opaque components. The averaged intensity conservation, eq. (2.32), simplifies to

$$\hat{\mathbf{s}} \cdot \nabla I_i(\mathbf{x}, \hat{\mathbf{s}}) = -\beta_i I_i(\mathbf{x}, \hat{\mathbf{s}}) + \kappa_{\mathrm{d},i} I_{\mathrm{b},i}(\mathbf{x}, \hat{\mathbf{s}}) + \frac{\sigma_{\mathrm{s},ii}}{4\pi} \int_{\Omega_i=0}^{4\pi} I_i(\mathbf{x}, \hat{\mathbf{s}}_i) \Phi_{ii}(\hat{\mathbf{s}}_i, \hat{\mathbf{s}}) \, \mathrm{d}\Omega_i$$

$$+ \frac{\sigma_{\mathrm{s},ji}}{4\pi} \int_{\Omega_i=0}^{4\pi} I_j(\mathbf{x}, \hat{\mathbf{s}}_i) \Phi_{ji}(\hat{\mathbf{s}}_i, \hat{\mathbf{s}}) \, \mathrm{d}\Omega_i, \qquad i,j=1,2; i \neq j, \qquad (2.33)$$

for a multi-phase media compose of two semitransparent phases. The averaged intensity conservation, eq. (2.32), simplifies to

$$\hat{\mathbf{s}} \cdot \nabla I_i(\mathbf{x}, \hat{\mathbf{s}}) = -\beta_i I_i(\mathbf{x}, \hat{\mathbf{s}}) + \kappa_{\mathrm{d},i} I_{\mathrm{b},i}(\mathbf{x}, \hat{\mathbf{s}}) + \frac{\sigma_{\mathrm{s},ii}}{4\pi} \int_{\Omega_i=0}^{4\pi} I_i(\mathbf{x}, \hat{\mathbf{s}}_i) \Phi_{ii}(\hat{\mathbf{s}}_i, \hat{\mathbf{s}}) \, \mathrm{d}\Omega_i,$$

$$i=1, \qquad (2.34)$$

for a multi-phase media composed of a semitransparent phases and an opaque phase. Therefore,

$$\sigma_{\mathrm{s},ii} = \sigma_{\mathrm{s,d},i} + \sigma_{\mathrm{s,refl},i}, \qquad (2.35)$$

$$\Phi_{ii} = \sigma_{\mathrm{s},ii}^{-1} \left(\Phi_{\mathrm{refl},i} \sigma_{\mathrm{s,refl},i} + \Phi_{\mathrm{d},i} \sigma_{\mathrm{s,d},i}\right), \qquad (2.36)$$

and for the two-phase media of semitransparent phases

$$\beta_i = \kappa_{\mathrm{d},i} + \sigma_{\mathrm{s},ii} + \sigma_{\mathrm{s},ji} = \beta_{\mathrm{d},i} + \sigma_{\mathrm{s,refl},i} + \sigma_{\mathrm{s},ji}, \qquad (2.37)$$

2.2. Heat transfer

while for the semitransparent/opaque two-phase media

$$\beta_i = \kappa_{\text{d},i} + \sigma_{\text{s},ii} + \kappa_{ij} = \beta_{\text{d},i} + \sigma_{\text{s,refl},i} + \kappa_{ij}. \tag{2.38}$$

Φ_{ii} and $\sigma_{\text{s},ii}$ are the scattering phase function and scattering coefficient associated with boundary reflection and internal scattering within phase i, Φ_{ij} and $\sigma_{\text{s},ij}$, $i \neq j$, are those associated with radiation leaving phase i and entering phase j, and κ_{ij} is associated with absorption of radiation from phase i at the boundary to phase j.

Equations (2.26) to (2.31) provide the mathematical basis for development of numerical techniques for the determination of continuum-scale radiative properties utilizing the exact geometry of multi-component media. They require the knowledge of the complete actual and blackbody discrete-scale radiative intensity fields in each component obtained for a selected model problem.

Once the continuum-scale radiative properties are known for a given medium, eqs. (2.32) can be solved for prescribed boundary conditions by using standard RTE solution techniques. Although the discrete-scale radiative intensities are no longer needed to solve the continuum-scale RTEs, they are still required to formulate the continuum-scale boundary conditions as described in the next section.

Continuum-scale boundary conditions

Equations (2.32) are subject to boundary conditions at the wall-medium interface at $\hat{\mathbf{s}} \cdot \hat{\mathbf{n}}_{\text{w}} > 0$, where $\hat{\mathbf{n}}_{\text{w}}$ is a unit normal vector pointing from the wall into the medium. The wall is assumed to consist of only a single component that can be either semi-transparent or opaque. The discrete-scale boundary conditions at the boundary of the multi-component medium are formulated analogously to the boundary conditions (2.17a) and (2.17b). They read for the semi-transparent and opaque walls, respectively:

$$L_i\left(\mathbf{r}_{iw}, \hat{\mathbf{s}}\right) = \int_{\Omega_{\hat{\mathbf{s}}_i \cdot \hat{\mathbf{n}}_{wi} > 0}} \tau''_{wi}\left(\mathbf{r}_{iw}, \hat{\mathbf{s}}_i, \hat{\mathbf{s}}\right) L_w\left(\mathbf{r}_{iw}, \hat{\mathbf{s}}_i\right) \hat{\mathbf{s}}_i \cdot \hat{\mathbf{n}}_{wi}\, d\Omega_i$$

$$- \int_{\Omega_{\hat{\mathbf{s}}_i \cdot \hat{\mathbf{n}}_{wi} < 0}} \rho''_{r,iw}\left(\mathbf{r}_{iw}, \hat{\mathbf{s}}_i, \hat{\mathbf{s}}\right) L_i\left(\mathbf{r}_{iw}, \hat{\mathbf{s}}_i\right) \hat{\mathbf{s}}_i \cdot \hat{\mathbf{n}}_{wi}\, d\Omega_i, \tag{2.39a}$$

$$L_i\left(\mathbf{r}_{iw}, \hat{\mathbf{s}}\right) = \varepsilon'_{r,wi}\left(\mathbf{r}_{iw}, \hat{\mathbf{s}}\right) L_{b,w}\left(\mathbf{r}_{iw}\right) - \int_{\Omega_{\hat{\mathbf{s}}_i \cdot \hat{\mathbf{n}}_{wi} < 0}} \rho''_{r,iw}\left(\mathbf{r}_{iw}, \hat{\mathbf{s}}_i, \hat{\mathbf{s}}\right) L_i\left(\mathbf{r}_{iw}, \hat{\mathbf{s}}_i\right) \hat{\mathbf{s}}_i \cdot \hat{\mathbf{n}}_{wi}\, d\Omega_i.$$

$$\tag{2.39b}$$

The variation of the discrete-scale radiative properties and the curvature of the wall–medium interface are assumed to be negligible over the interface area associated with the averaging volume V adjacent to said boundary, $A_{iw,\hat{s}\cdot\hat{n}_{wi}>0}$. The continuum-scale boundary conditions are obtained by surface averaging of the boundary intensity $L_i(\mathbf{r}_{iw},\hat{\mathbf{s}})$,

$$I_i(\mathbf{x}_w,\hat{\mathbf{s}}) \frac{\int_{A_{iw,\hat{s}\cdot\hat{n}_{wi}>0}} L_i(\mathbf{r}_{iw},\hat{\mathbf{s}})\,\mathrm{d}A}{\int_{A_{w,\hat{s}\cdot\hat{n}_{w}>0}} \mathrm{d}A}, \qquad (2.40)$$

where A_w is the portion of the wall-medium interface inside the averaging volume V adjacent to the wall. Applying eq. (2.40) to eqs. (2.39a) and (2.39b), and interchanging the order of integration with respect to Ω_i and A on the right-hand side of the resulting equation, leads to:

$$I_i(\mathbf{x}_w,\hat{\mathbf{s}}) = \int_{\Omega_{\hat{s}_i\cdot\hat{n}_{wi}>0}} \tau''_{wi}(\mathbf{x}_w,\hat{\mathbf{s}}_i,\hat{\mathbf{s}}) I_{wi}(\mathbf{x}_w,\hat{\mathbf{s}}_i)\,\hat{\mathbf{s}}_i\cdot\hat{\mathbf{n}}_{wi}\,\mathrm{d}\Omega_i$$
$$- \int_{\Omega_{\hat{s}_i\cdot\hat{n}_{wi}<0}} \rho''_{r,iw}(\mathbf{x}_w,\hat{\mathbf{s}}_i,\hat{\mathbf{s}}) I_i(\mathbf{x}_w,\hat{\mathbf{s}}_i)\,\hat{\mathbf{s}}_i\cdot\hat{\mathbf{n}}_{wi}\,\mathrm{d}\Omega_i, \qquad (2.41\mathrm{a})$$

$$I_i(\mathbf{x}_w,\hat{\mathbf{s}}) = \varepsilon'_{r,wi}(\mathbf{x}_w,\hat{\mathbf{s}}) I_{b,wi}(\mathbf{x}_w,\hat{\mathbf{s}}) - \int_{\Omega_{\hat{s}_i\cdot\hat{n}_{wi}<0}} \rho''_{r,iw}(\mathbf{x}_w,\hat{\mathbf{s}}_i,\hat{\mathbf{s}}) I_i(\mathbf{x}_w,\hat{\mathbf{s}}_i)\,\hat{\mathbf{s}}_i\cdot\hat{\mathbf{n}}_{wi}\,\mathrm{d}\Omega_i. \qquad (2.41\mathrm{b})$$

Comparison of discrete-scale, eqs. (2.16) and (2.17), versus continuum-scale, eqs. (2.32) and (2.41), radiative transfer calculations in opaque and semitransparent porous media shows good agreement [161] within the local statistical variations of the discrete-scale geometry. In addition, the continuum-scale approach offers significant reduction in computational expense.

Volume-averaged divergence of radiative flux

The continuum-scale divergences of the radiative fluxes are obtained by integrating eqs. (2.32) over all wavelengths and solid angles resulting in

$$\nabla\cdot\langle\mathbf{q}''_{r,i}\rangle = \int_{\lambda=0}^{\infty} \left\{ 4\pi\left[\kappa_{\lambda,d,i} I_{\lambda b,i}(\mathbf{x},\hat{\mathbf{s}}) + \sum_{j=N_{i,1}+1}^{N_i} \kappa_{\lambda,ji} I_{\lambda b,j}(\mathbf{x},\hat{\mathbf{s}})\right]\right.$$
$$- \left[\kappa_{\lambda,d,i} + \sum_{j=N_{i,1}+1}^{N_i}\kappa_{\lambda,ij} + \sum_{j=1}^{N_{i,1}}\sigma_{\lambda,s,ij}\right]\int_{\Omega=0}^{4\pi} I_{\lambda,i}(\mathbf{x},\hat{\mathbf{s}})\,\mathrm{d}\Omega$$
$$\left.+ \sum_{j=1}^{N_{i,1}}\sigma_{\lambda,sji}\int_{\Omega=0}^{4\pi} I_{\lambda,j}(\mathbf{x},\hat{\mathbf{s}})\,\mathrm{d}\Omega\right\}\mathrm{d}\lambda, \qquad i=1,\ldots,M_1, \qquad (2.42)$$

2.2. Heat transfer

where subscript λ is reintroduced. The continuum-scale divergences are used in the averaged energy equations, e.g., for a two-phase medium described by eqs. (2.10) and (2.11). Alternatively, the average of the divergence of the radiative flux is given by averaging the discrete-scale divergence of the radiative flux, which is obtained by integrating eq. (2.12) over all wavelengths and solid angles, to yield

$$\langle \nabla \cdot \mathbf{q}''_{\mathrm{r},i} \rangle = \int_{\lambda=0}^{\infty} \kappa_{\lambda,\mathrm{d},i} \left\{ 4\pi I_{\lambda \mathrm{b},i}(\mathbf{x},\hat{\mathbf{s}}) + \int_{\Omega=0}^{4\pi} I_{\lambda,i}(\mathbf{x},\hat{\mathbf{s}})\,\mathrm{d}\Omega \right\} \mathrm{d}\lambda, \qquad (2.43)$$

by assuming that the variation of the discrete-scale radiative properties is small enough so that the corresponding variation of the discrete scale radiative properties inside the averaging volume V becomes negligible [119]. This approach is often used in studies where radiation-turbulence interactions are of interest and therefore the time averaged divergence of the radiative flux is needed [35, 85].

2.2.2 Conduction

The two divergence terms present in eqs. (2.10) and (2.11),

$$\nabla \cdot (k_i \nabla \langle T_i \rangle) - \nabla \cdot \left[k_i \frac{1}{V} \int_{A_{ij}} T_i \cdot \hat{\mathbf{n}}_{ji}\,\mathrm{d}A \right], \qquad (2.44)$$

account for conduction in the phases.

Two approaches are generally used in literature to close these terms: One-equation models, which assume local thermal equilibrium, and two-equation models, which assume local thermal equilibrium not to be valid [175, 233, 102]. Local thermal equilibrium assumes

$$\langle T_{\mathrm{f}} \rangle^{\mathrm{f}} = \langle T_{\mathrm{s}} \rangle^{\mathrm{s}} = \langle T \rangle \qquad (2.45)$$

to be a reasonable approximation. Therefore, adding eqs. (2.10) and (2.11), and using the simplification

$$\frac{1}{V} \int_{A_{ij}} \langle T_i \rangle^i \hat{\mathbf{n}}_{ji}\,\mathrm{d}A \approx \langle T_i \rangle^i \nabla f_{v,i}, \qquad (2.46)$$

results in

$$\begin{aligned}
& [f_{v,\mathrm{f}}(\rho c_p)_{\mathrm{f}} + f_{v,\mathrm{s}}(\rho c_p)_{\mathrm{s}}] \frac{\partial \langle T \rangle}{\partial t} \\
& = \nabla \cdot \left[(f_{v,\mathrm{f}} k_{\mathrm{f}} + f_{v,\mathrm{s}} k_{\mathrm{s}}) \nabla \langle T \rangle - k_{\mathrm{f}} \frac{1}{V} \int_{A_{\mathrm{sf}}} \tilde{T}_{\mathrm{f}} \cdot \hat{\mathbf{n}}_{\mathrm{sf}}\,\mathrm{d}A + k_{\mathrm{s}} \frac{1}{V} \int_{A_{\mathrm{sf}}} \tilde{T}_{\mathrm{s}} \cdot \hat{\mathbf{n}}_{\mathrm{sf}}\,\mathrm{d}A \right] \\
& = \nabla \cdot (k_{\mathrm{e}} \nabla \langle T \rangle) = \nabla \cdot (k_{\mathrm{e}} \nabla \langle T_i \rangle^i).
\end{aligned} \qquad (2.47)$$

The additional source terms present in eqs. (2.10) and (2.11) are neglected for brevity. Equation (2.47) is therefore closed by introducing k_e.

If local thermal equilibrium is not valid, it has been shown [175] that the conduction term is expressed by

$$\nabla \cdot \left(k_{e,i} \nabla \langle T_i \rangle^i + k_{e,ij} \nabla \langle T_j \rangle^j \right) , \tag{2.48}$$

where $k_{e,ij} = k_{e,ji}$.

2.2.3 Convection

The heat flux terms integrated over the phase boundaries, present in eqs. (2.10) and (2.11), describe the interfacial heat fluxes originating from conductive and radiative heat fluxes over the phase boundaries:

$$\frac{1}{V} \int_{A_{sf}} k_f \nabla T_f \cdot \hat{\mathbf{n}}_{sf} dA - \frac{1}{V} \int_{A_{sf}} \mathbf{q}''_{r,f} \cdot \hat{\mathbf{n}}_{sf} dA = \frac{1}{V} \int_{A_{sf}} \mathbf{q}''_{tot,f} \cdot \hat{\mathbf{n}}_{sf} dA$$

$$= \frac{1}{V} \int_{A_{sf}} k_s \nabla T_s \cdot \hat{\mathbf{n}}_{sf} dA - \frac{1}{V} \int_{A_{sf}} \mathbf{q}''_{r,s} \cdot \hat{\mathbf{n}}_{sf} dA = \frac{1}{V} \int_{A_{sf}} \mathbf{q}''_{tot,s} \cdot \hat{\mathbf{n}}_{sf} dA$$

$$= h_{sf} A_0 (\langle T_s \rangle^s - \langle T_f \rangle^f). \tag{2.49}$$

The heat transfer coefficient, h_{sf}, describes the convective heat exchange between the phases in the porous media. Often the pure convective heat transfer coefficient is of interest, as the radiative flux at the boundary depends on the averaging procedure and the discrete-scale radiative boundary conditions. Therefore the heat transfer coefficient of pure convection is given by

$$\frac{1}{V} \int_{A_{sf}} k_f \nabla T_f \cdot \hat{\mathbf{n}}_{sf} dA = \frac{1}{V} \int_{A_{sf}} k_s \nabla T_s \cdot \hat{\mathbf{n}}_{sf} dA = \frac{1}{V} \int_{A_{sf}} \mathbf{q}''_{conv} \cdot \hat{\mathbf{n}}_{sf} dA$$

$$= h_{sf} A_0 (\langle T_s \rangle^s - \langle T_f \rangle^f). \tag{2.50}$$

2.2.4 Equation of state

The equation of state, for an ideal gas described by

$$p_i M = R \rho_i T_i , \tag{2.51}$$

is averaged to yield

$$\langle p_i \rangle^i M = R \langle \rho_i \rangle^i \langle T_i \rangle^i . \tag{2.52}$$

2.3 Mass transfer

Mass transfer of a two-phase medium, e.g., a connected moving fluid phase and a connected stationary semi-transparent rigid solid phase, are described by mass

2.3. Mass transfer

conservation,
$$\frac{\partial \rho_i}{\partial t} + \nabla \cdot (\rho_i \mathbf{u}_i) = S_{\mathrm{m},i}, \tag{2.53}$$

species conservation, given for the molar concentration c_{ki} of species k in the (fluid) phase i,
$$\frac{\partial c_{ki}}{\partial t} + \nabla \cdot (c_{ki} \mathbf{u}_i) = \nabla \cdot (\mathbf{D}_i \nabla c_{ki}), \tag{2.54}$$

and momentum conservation in the Newtonian incompressible fluid phase, neglecting gravity forces,
$$\rho_i \left(\frac{\partial \mathbf{u}_i}{\partial t} + \mathbf{u}_i \cdot \nabla \mathbf{u}_i \right) = -\nabla p_i + \mu_i \Delta \mathbf{u}_i. \tag{2.55}$$

No slip at the fluid-solid phase boundary as well as passive dispersion and impermeable boundaries are assumed.

Volume averaging of equation (2.53) results in
$$f_{v,i} \frac{\partial \langle \rho_i \rangle^i}{\partial t} + \nabla \cdot \left(f_{v,i} \langle \rho_i \rangle^i \langle \mathbf{u}_i \rangle^i \right) + \nabla \cdot \langle \tilde{\rho}_i \tilde{\mathbf{u}}_i \rangle = \langle S_{\mathrm{m},i} \rangle. \tag{2.56}$$

For steady, incompressible flow without mass sources, the relation simplifies to
$$\nabla \cdot \langle \mathbf{u}_i \rangle = 0. \tag{2.57}$$

Volume averaging of equation (2.54), by using the simplification given in eq. (2.46) for c_{ji} instead of T_i, results in

$$\begin{aligned}
f_{v,i} \frac{\partial \langle c_{ki} \rangle^i}{\partial t} &+ \nabla \cdot \left(f_{v,i} \langle \mathbf{u}_i \rangle^i \langle c_{ki} \rangle^i \right) \\
&= \nabla \cdot \left(\mathbf{D}_i f_{v,i} \nabla \langle c_{ki} \rangle^i + \mathbf{D}_i \frac{1}{V} \int_{A_{ij}} \tilde{c}_{ki} \hat{\mathbf{n}}_{ji} \mathrm{d}A - \langle \tilde{\mathbf{u}}_i \tilde{c}_{ki} \rangle \right) \\
&= \nabla \cdot \left(f_{v,i} \mathbf{D}_{\mathrm{e},i} \cdot \nabla \langle c_{ki} \rangle^i \right).
\end{aligned} \tag{2.58}$$

Volume averaging of equation (2.55), by using the simplification given in eq. (2.46) for p_i and $\mathbf{u}_i$ instead of T_i, results in

$$f_{v,i} \rho_i \frac{\partial \langle \mathbf{u}_i \rangle^i}{\partial t} + f_{v,i} \rho_i \langle \mathbf{u}_i \rangle^i \nabla \cdot \langle \mathbf{u}_i \rangle^i = - f_{v,i} \nabla \langle p_i \rangle^i + f_{v,i} \mu_i \Delta \langle \mathbf{u}_i \rangle^i$$
$$+ \frac{1}{V} \int_{A_{ij}} (\tilde{p}_i - \mu_i \nabla \tilde{\mathbf{u}}_i) \hat{\mathbf{n}}_{ji} \mathrm{d}A - \rho_i \nabla \langle \tilde{\mathbf{u}}_i \tilde{\mathbf{u}}_i \rangle. \tag{2.59}$$

It has been shown [232] that the inertia effects of the averaged momentum equation can be neglected (unlike on the discrete-scale), therefore eq. (2.59) simplifies to

$$\begin{aligned}0 &= -\nabla \langle p_i \rangle^i + \mu_i \Delta \langle \mathbf{u}_i \rangle^i + \frac{1}{V_i} \int_{A_{ij}} (\tilde{p}_i - \mu_i \nabla \tilde{\mathbf{u}}_i) \, \hat{\mathbf{n}}_{ji} \mathrm{d}A \\ &= -\nabla \langle p_i \rangle^i + \mu_i \Delta \langle \mathbf{u}_i \rangle^i + \mu_i \left(-\frac{f_{v,i}}{K} - \frac{F_{\mathrm{DF}} f_{v,i}^2}{\nu} \langle \mathbf{u}_i \rangle^i \right) \langle \mathbf{u}_i \rangle^i \ . \end{aligned} \quad (2.60)$$

Length scale constrains show that the Brinkman correction term, $\mu_i \Delta \langle \mathbf{u}_i \rangle^i$, can be neglected for cases where no significant boundary effects are present. Therefore eq. (2.60) reduces to the well know Darcy's equation with Dupuit-Forchheimer extension

$$0 = -\nabla \langle p_i \rangle^i - \frac{\mu_i}{K} \langle \mathbf{u}_i \rangle - \rho_i F_{\mathrm{DF}} |\langle \mathbf{u}_i \rangle| \langle \mathbf{u}_i \rangle \ . \quad (2.61)$$

Chapter 3
Methodology

In this chapter, the methodologies with which the effective heat and mass transfer properties are determined are introduced, namely extinction coefficient, scattering coefficient, scattering phase function, conductivity, heat transfer coefficient, permeability, Dupuit-Forchheimer coefficient, and dispersion tensor. Their calculation is based on direct numerical pore-level (discrete-scale) simulations (DPLS, also called explicit numerical simulations) on the exact geometry, obtained by CT, and is partially introduced in [160].

For radiative characterization, the radiative distribution functions [211] are introduced and determined by collision-based Monte Carlo (MC). The extraction of the radiative properties is described and comparison to a direct approach [119, 161] is done. Stationary conduction for multiple phases in a quasi 1D situation is solved by the finite volume (FV) technique, and subsequently extraction of the effective heat conductivity is introduced that is valid for pure conduction in a multi-phase media with stagnant fluid [175, 163]. Mass, momentum and energy conservation in the incompressible, laminar fluid phase are calculated by FV technique and used for the determination of: effective heat transfer coefficient dependent on pure convection, permeability, Dupuit-Forchheimer coefficient, and dispersion tensor. Additionally, the latter approach is used for the determination of tortuosity and residence time distributions of the the multi-phase media.

3.1 Heat transfer

3.1.1 Radiation

The methodology to determine the effective radiative properties of two-phase media is outlined below. The most general case of a medium consisting of two semitransparent phases is described, and then simplification for a medium composed of a transparent and an opaque phase are given.

Geometric optics is assumed. This assumption is true when the characteristic size parameters $\xi = \pi d/\lambda \gg 1$ for both phases [141]. This is fulfilled in the cases investigated in this book as the pore/particles are in the µm range and the wavelengths of interest in the ultra violet (UV), visible and near-infrared (IR) region.

Equation (2.33) is recalled. The continuum scale radiative properties, defined by eqs. (2.26) to (2.31), and reintroduced in eqs. (2.35) to (2.38) for a two-phase media of semitransparent phases, need to be determined.

An approach based on the differential interpretation of the continuum-scale radiative properties is presented in [120, 161]. In this book, the methodology presented in [211], which is based on the integral interpretation of the radiative properties and calculated based on radiative distribution functions (RDFs), has been chosen. The two approaches are equal when the RDFs are calculated on sufficiently short path lengths, shown as an example for the calculated scattering phase functions in a packed bed of semitransparent particles with path lengths of 500 µm in figure 3.1.

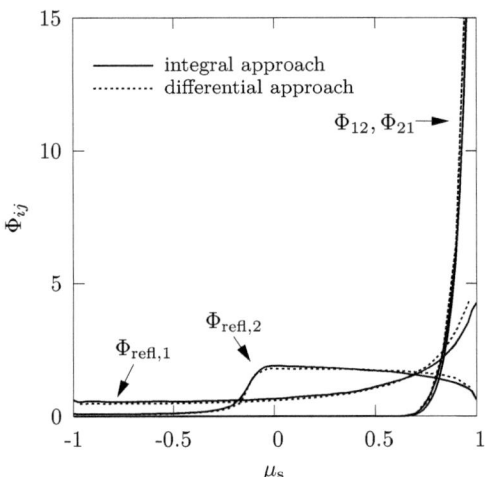

Figure 3.1: Scattering phase functions calculated based on the integral approach [211] or the differential approach [120, 161].

The collision-based MC ray-tracing method is used [52]. A large number of stochastic rays are launched within a representative elementary volume (REV). The rays are emitted isotropically and are uniformly distributed over REV. They undergo scattering/absorption internally and reflection/refraction at the

3.1. Heat transfer

fluid-solid interface. The distance between emission and collision points and the direction of incidence at the interface are recorded for each ray.

The calculations are performed on the tomographic data where the intersection point between a ray and the solid-fluid interface is found by following the ray in discrete steps until the boundary is reached. Finally, the exact determination is achieved by applying the bisection method. Note that the interface is described by a continuous function and that no numerical grid is required. The computations are performed with an in-house Fortran 95 code.

The following probability density and cumulative distribution functions and the corresponding medium properties are then computed. For the attenuation (extinction) path length within phase i,

$$F_{e,i}(s) = \frac{1}{N_{\text{ray},i}} \sum_{k=1}^{\text{ray},i} \delta(s - s_k), \qquad i = 1, 2, \tag{3.1}$$

$$G_{e,i}(s) = \int_0^s F_{e,i}(s^*) \mathrm{d}s^* \approx 1 - \exp(-\beta_i s), \qquad i = 1, 2. \tag{3.2}$$

For the absorption path length within phase i,

$$F_{a,i}(s) = \frac{1}{N_{\text{ray},i}} \sum_{k=1}^{\text{ray},i} \delta(s - s_{a,k}), \qquad i = 1, 2, \tag{3.3}$$

$$G_{a,i}(s) = \int_0^s F_{a,i}(s^*) \mathrm{d}s^* \approx \frac{\kappa_i}{\beta_i} \left(1 - \exp(-\beta_i s)\right), \qquad i = 1, 2. \tag{3.4}$$

For the scattering path length associated with reflection and internal scattering within phase i, and refraction from phases i to j,

$$F_{s,ij}(s) = \frac{1}{N_{\text{ray},i}} \sum_{k=1}^{\text{ray},i} \delta(s - s_{s,k}) H(\rho_r'' - \Re), \qquad i = j = 1, 2, \tag{3.5}$$

$$F_{s,ij}(s) = \frac{1}{N_{\text{ray},i}} \sum_{k=1}^{\text{ray},i} \delta(s - s_{s,k}) H(\Re - \rho_r''), \qquad i, j = 1, 2; i \neq j, \tag{3.6}$$

$$G_{s,ij}(s) = \int_0^s F_{s,ij}(s^*) \mathrm{d}s^* \approx \frac{\sigma_{ij}}{\beta_i} \left(1 - \exp(-\beta_i s)\right), \qquad i, j = 1, 2, \tag{3.7}$$

with the random number $\Re$ and the Heaviside step function H (which adopts the convention $H(0) = 1$ instead of the conventional definition $H(0) = 0.5$). For the direction of incidence at the phase interface within phase i,

$$F_{\mu_{\text{in}},i}(\mu_{\text{in}}) = \frac{1}{N_{\text{att},i}} \sum_{k=1}^{\text{att},i} \delta(\mu_{\text{in}} - \mu_{\text{in},k}), \qquad i = 1, 2, \tag{3.8}$$

$$\Phi_{ij}(\mu_\text{s}) \approx \int_{\mu_\text{in}=0}^{1}\int_{\varphi_\text{d}=0}^{\pi}\int_{\mu_\text{refl}=0}^{1} \delta(\mu_\text{s}-\sqrt{(1-\mu_\text{in}^2)(1-\mu_\text{refl}^2)}\cos\varphi_\text{d}-\mu_\text{in}\mu_\text{refl})$$
$$\times \rho_\text{r}''(\mu_\text{in},\mu_\text{refl},\varphi_\text{d})F_{\mu_\text{in},i}\mu_\text{refl}d\mu_\text{refl}d\varphi_\text{d}d\mu_\text{in}$$
$$\bigg\backslash \int_{\mu_\text{in}=0}^{1}\int_{\varphi_\text{d}=0}^{\pi}\int_{\mu_\text{refl}=0}^{1} \rho_\text{r}''(\mu_\text{in},\mu_\text{refl},\varphi_\text{d})F_{\mu_\text{in},i}\mu_\text{refl}d\mu_\text{refl}d\varphi_\text{d}d\mu_\text{in}, \, i=j=1,2\,,$$
(3.9)

$$\Phi_{ij}(\mu_\text{s}) \approx \int_{\mu_\text{in}=0}^{1}\int_{\varphi_\text{d}=0}^{\pi}\int_{\mu_\text{t}=0}^{1} \delta(\mu_\text{s}-\sqrt{(1-\mu_\text{in}^2)(1-\mu_\text{t}^2)}\cos\varphi_\text{d}-\mu_\text{in}\mu_\text{t})$$
$$\times (1-\rho_\text{r}''(\mu_\text{in},\mu_\text{t},\varphi_\text{d}))F_{\mu_\text{in},i}\mu_\text{t}d\mu_\text{t}d\varphi_\text{d}d\mu_\text{in}$$
$$\bigg\backslash \int_{\mu_\text{in}=0}^{1}\int_{\varphi_\text{d}=0}^{\pi}\int_{\mu_\text{t}=0}^{1} (1-\rho_\text{r}''(\mu_\text{in},\mu_\text{t},\varphi_\text{d}))F_{\mu_\text{in},i}\mu_\text{t}d\mu_\text{t}d\varphi_\text{d}d\mu_\text{in}, \, i,j=1,2; i\neq j\,.$$
(3.10)

In eqs. (3.1) to (3.10), $N_{\text{ray},i}$ is the number of rays launched in phase i within REV, and $N_{\text{att},i}$ is the number of rays attenuated in phase i at the phase interface. μ_in, μ_refl, μ_t, and μ_s are the cosine of incidence, reflection, transmission, and scattering angle, respectively. φ_d is the difference of incidence and reflection azimuthal angles, and ρ_r'' describes the bi-directional reflectivity at the solid-fluid boundary.

No intensity exists in the opaque phase for a two-phase medium consisting of a semitransparent and an opaque phase. Therefore, $i=1$ and only one distribution function of extinction, absorption and scattering path length and only one distribution function of cosine of interface incidence needs to be calculated. For such a case the methodology has been applied in [164] to a SiC foam.

3.1.2 Conduction

The one-equation averaging model, given by eq. (2.47), is valid if local thermal equilibrium is a reasonable assumption. This is the case if one of the following conditions occurs: (i) $f_{v,\text{f}}$ or $f_{v,\text{s}}$ tends towards zero, (ii) the difference in the physical properties of the two phases tends towards zero, and (iii) the ratio of length scales $(d/l_\text{system})^2$ tends towards zero [233]. For reticulate foam structures, condition (i) occurs, while for packed beds condition (iii) occurs.

Equation (2.47) results in a linear temperature distribution for a 1D case. The heat rate per unit area is then given by

$$\langle q''\rangle = k_\text{e}\frac{T_1-T_2}{l}\,.$$
(3.11)

A quasi one-dimensional situation on the discrete-scale is created by solving eqs. (2.8) and (2.9) for steady-state and neglecting radiative and other source terms

3.1. Heat transfer

in a cubic porous sample. The boundary conditions for the lateral walls are given by

$$\mathbf{q}'' \cdot \hat{\mathbf{n}} = 0, \tag{3.12}$$

for the solid-fluid interface by

$$T_f = T_s, \hat{\mathbf{n}} \cdot k_s \nabla T_s = \hat{\mathbf{n}} \cdot k_f \nabla T_f, \tag{3.13}$$

at the inlet by

$$T_s = T_f = T_1, \tag{3.14}$$

and at the outlet by

$$T_s = T_f = T_2. \tag{3.15}$$

The heat flux across the sample at any given cross-sectional plane perpendicular to the main heat flow direction is then given by

$$\langle q'' \rangle = \frac{-\int_{A_s} k_s \nabla T_s \cdot \hat{\mathbf{n}} \mathrm{d}A_s - \int_{A_f} k_f \nabla T_f \cdot \hat{\mathbf{n}} \mathrm{d}A_f}{A_s + A_f}. \tag{3.16}$$

Combining eqs. (3.11) and (3.16) results in the determination of the effective conductivity in a porous media by

$$k_e = l \frac{-\int_{A_s} k_s \nabla T_s \cdot \hat{\mathbf{n}} \mathrm{d}A_s - \int_{A_f} k_f \nabla T_f \cdot \hat{\mathbf{n}} \mathrm{d}A_f}{(T_1 - T_2)(A_s + A_f)}. \tag{3.17}$$

The FV technique with successive over-relaxation is used to solve the governing equations. The numerical methodology is described in [172] and was previously used in [163].

3.1.3 Convection

Heat transfer between the void and solid phase in a porous media is described by the heat transfer coefficient. It is derived for the volume-averaged energy conservation equation and given by eq. (2.50).

Finite volume technique is used [91] to derive the heat transfer coefficient for a constant surface temperature. Mass, momentum and energy conservation, described by eqs. (2.53), (2.55), and (2.8), are solved for steady-state in a incompressible laminar fluid phase on the discrete-scale. The domain used for the calculations is composed of a square duct containing a sample of the multi-phase media and an undisturbed inlet and outlet region, depicted in figure 3.2. The boundary conditions at the inlet is given by

$$\mathbf{u} \cdot \hat{\mathbf{n}} = -u_{\mathrm{in}}, T = T_{\mathrm{in}}, \tag{3.18}$$

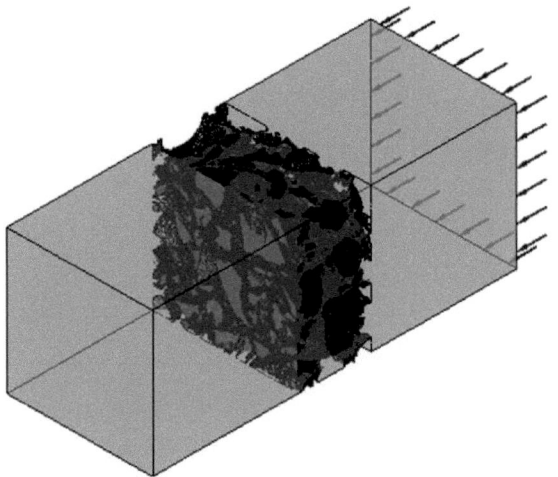

Figure 3.2: Schematic of the domain, consisting of a square duct containing a sample of the multi-phase media, an inlet and outlet regions.

at the lateral walls by

$$\mathbf{u} \cdot \hat{\mathbf{n}} = 0\,, \hat{\mathbf{n}} \cdot \nabla \mathbf{u} = 0\,, \mathbf{q}'' \cdot \hat{\mathbf{n}} = 0\,, \tag{3.19}$$

at the solid-fluid interface by

$$\mathbf{u} = 0\,, T = T_{\text{sf}}\,, \tag{3.20}$$

and at the outlet by

$$p = p_{\text{atm}}\,. \tag{3.21}$$

A mesh, generated with an in-house mesh generator for unstructured body-fitted grids, is used for the calculations [57, 58]. The mesh generator covers the domain by tetrahedral elements and subsequently refines the elements at the phase boundary. Finally, a rounding, cutting, and smoothing process is carried out to achieve an accurate domain representation.

The calculated velocity and temperature fields and the boundary heat fluxes are used to derive the heat transfer coefficient by

$$h_{\text{sf}} = \frac{\int_{A_{\text{sf}}} q'' \text{d} A_{\text{sf}}}{A_{\text{sf}} \Delta T_{\text{lm}}}\,. \tag{3.22}$$

The heat transfer coefficient and Nu, respectively, are calculated for different Re and Pr numbers and fitted to relations of the form

$$\text{Nu} = a_1 + a_2 \text{Re}^{a_3} \text{Pr}^{a_4}\,. \tag{3.23}$$

3.2 Mass transfer

3.2.1 Permeability and Dupuit-Forchheimer coefficient

Pressure and velocity fields through multi-phase media are influenced by permeability, K, and for higher Re numbers by an inertia-induced term described by the Dupuit-Forchheimer coefficient, F_{DF}. They are determined by solving mass and momentum conservation, described by eqs. (2.53) and (2.55), in the incompressible, laminar fluid phase with the same boundary conditions and on the same domain as described in section 3.1.3. The calculated velocity and pressure fields are then used in the averaged momentum equation, described by eq. (2.61). This equation can be normalized to explicitly show the linear dependence of the normalized pressure gradient, Π_{pg}, on the Re number (in 1D):

$$\frac{\nabla \langle p_f \rangle^f d^2}{\mu} = \Pi_{pg} = -\frac{d^2}{K} - F_{DF} d \text{Re} = -a_1 - a_2 \text{Re}. \tag{3.24}$$

Therefore Re dependent determination of the pressure gradient over the porous medium allows for the determination of permeability and Dupuit-Forchheimer coefficient.

This methodology has been used for the derivation of the flow characteristics through a SiC foam [162].

3.2.2 Tortuosity and residence time distributions

Tortuosity is defined as the ratio of the real length of the connected pore channels to the thickness of the porous sample in the main flow direction,

$$\tau = \frac{l_{path}}{l_{sample}}. \tag{3.25}$$

The residence time required for a particle to flow through the porous sample is defined as

$$t = \int_{l_{path}} \frac{1}{|\mathbf{u}|} dl. \tag{3.26}$$

The calculated velocity distribution, obtained as described in section 3.2.1, is used to create a large number of stream lines uniformly distributed over the inlet for each Re number. Their length and the integrated time on the stream line are calculated and used for additional multi-phase medium characterization.

3.2.3 Dispersion

The formal solution of the species conservation equation on the continuum-scale, given by eq. (2.58), for a constant dispersion tensor in 1D reads

$$\langle c_{kf} \rangle^f = \frac{Q}{\sqrt{D_{e,x}\pi t}} \exp\left(-\frac{(x - x_0 - \langle u_{f,x}\rangle^f t)^2}{4D_{e,x}t}\right) = Q^* \exp\left(-\frac{\Delta x^2}{2S_{c,x}^2}\right). \quad (3.27)$$

Q is related to the total amount of dispersed medium initially induced at the origin (x_0, y_0, and z_0). The dispersion tensor in an isotropic medium can be decomposed in a axial $D_\parallel$ (parallel to the main flow direction) and a radial part $D_\perp$. D_x and D_y are equal and correspond to $D_\perp$ and D_z corresponds to $D_\parallel$ for a main flow along the z-direction. The formal solution links the dispersion tensor to the standard deviation of the normally distributed concentration at a specific time [189],

$$S_{c,i} = \sqrt{2D_{e,i}t}. \quad (3.28)$$

The calculated velocity distribution, obtained as described in section 3.2.1, is used to create a large number of stream lines uniformly distributed over the inlet for each Re number. The standard deviation in each direction is determined by calculating the spatial displacement at a specific time instant (e.g., in x-direction: $\Delta x = x(t) - x(t_0)$, or in z-direction: $\Delta z = z(t) - z(t_0) - u_0 \cdot t$) and fitting to a standard Gauss distribution.

For no molecular diffusion the Re dependence of the dispersion is described by

$$D_i = a_1 \text{Re}^{a_2}. \quad (3.29)$$

Chapter 4
Morphological characterization

The governing equations introduced in chapter 2 show that the effective transport properties strongly depend on the fluid-solid interface geometry, A_{sf}. Therefore an accurate representation of the fluid-solid boundary (morphology or discrete-scale geometry) is needed.

The progress in tomography techniques in recent decades has facilitated a broad range of their applications in science, engineering, and medicine. They are an excellent tool to provide accurate geometrical information of internal material structure. Computed tomography (CT) has been used to experimentally obtain the exact discrete-scale geometry to be incorporated in the subsequent discrete-scale numerical calculations of the morphological, effective transport and mechanical properties. Morphological properties such as porosity, specific surface area, pore- and particle-size distributions, of fibrous material [45], ceramic foams [105, 164], and packed beds [82] have been determined based on the exact geometry obtained by CT. Radiative characterization – including the determination of effective extinction and scattering coefficients and scattering phase functions – of foams and packed beds consisting of semi-transparent solid materials [238, 38, 37, 82] and of opaque solid material [165, 78, 36, 83] have been studied. The effective heat conductivity and heat transfer coefficient of foams [105, 163, 164] and packed beds [81] have been calculated employing CT. Flow properties such as permeability and Dupuit-Frochheimer coefficient of fibrous materials [45], foams [105, 162], packed beds [81], and rock [169] have been calculated. Further examples include the determination of sound absorption in foams [15] and mechanical properties, such as Poisson's ratio, bulk and Young's moduli, of foams [105], packed beds [127], rock formations [127], and biological structures [144, 127].

Additionally, tomography has been used for non-intrusive 3D field measurements [139] of temperature [218], density [104], concentration [190], refractive index [51], and absorption coefficient. These measurements allow for the validation of 3D numerical solutions.

This chapter specifies how the tomographic data is obtained and enhanced. Subsequent phase segmentation and digitalization of the CT data is described. The methodology for the numerical calculation of the morphological characteristics, namely porosity, specific surface area, pore- and particle-size distributions, and REV, are introduced, and experimental techniques for comparison are briefly described.

4.1 Computed tomography

Basically, CT uses multiple 2D projections of an object to reconstruct its complete 3D image or internal structure. Mathematical methods such as inverse Radon transform are used, in which the intensity data acquired by a detector at multiple tilt angles are used to determine the spatial (2D or 3D) distribution of the measured property. Attenuation, phase contrast or diffraction mode are used to inversely determine the distribution of the absorption coefficient, refractive index, and apparent scattering coefficient within the sample [88]. Phase contrast is specially suited for materials composed of components with similar absorption behavior. Figure 4.1 exemplarily shows absorption and phase contrast tomography of the same sample of animal tissue obtained with absorption-based and phase contrast-based CT at the TOMCAT beamline at the Swiss Light Source (SLS) of the Paul Scherrer Institute (PSI, in Villigen, Switzerland).

Tomography techniques are classified according to the physical mechanism used as a source of the 2D projections, determining the size and resolution of the scans. The resolution of a scan is described by the size of a voxel, a volumetric pixel which represents the smallest detectable subvolume in 3D Cartesian coordinates. The most common tomography techniques include the atom probe tomography, electron tomography, neutron tomography, X-ray tomography, and radio tomography. The length scales that can be resolved range from those typical for atoms and molecules (atom probe tomography) up to celestial objects (radio tomography). The majority of the tomography techniques are non-destructive, in which sectioning is done by collecting projections by a detector at multiple viewing directions. An example of a destructive tomography technique is the focused ion beam tomography used for micro- and nanoscale in-situ milling, in which the sectioning is done by cutting and etching.

The X-ray computed tomography is one of the most widely used techniques. It is based on two types of X-ray sources: micro-focus X-ray tube (microtomography) [207] and synchrotron radiation (synchrotron tomography) [208,

4.2. Segmentation and digitalization

(a) (b)

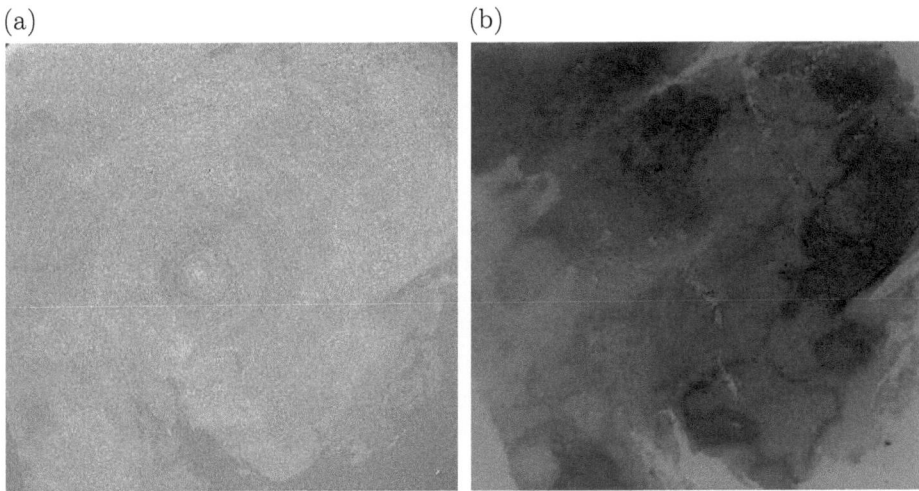

Figure 4.1: Absorption-based CT with voxel size 3.7 µm (a) and phase-based CT with voxel size 7.4 µm (b) of animal tissue material. The edge length of the samples is 5.1 mm.

200]. The resolution of the micro-tomography typically varies between 10 and 100 µm, while the one of the synchrotron tomography can be as low as a few hundred nanometers. Synchrotron radiation is non-diverging, coherent and has a high photon flux, typically 3 orders of magnitude higher than micro-tomography, allowing for high-quality tomograms within a comparably short period of time.

4.2 Segmentation and digitalization

The data obtained by CT consists of 2 byte (0–65535) optical density values, $\alpha(\mathbf{x})$, arranged on a 3D Cartesian grid. Multi-phase media are composed of different regions, each with a comparatively uniform optical density. These different regions, called phases, reflect the different components present within the material. In order to digitalize and determine the morphological, effective transport and mechanical properties of the material, these phases must be identified, thereby partitioning the grey-scale image into disjoint segments. Therefore, each voxel must be assigned to one phase. This process is called segmentation.

Prior to segmentation, the tomographic data needs to be enhanced by digital image processing. Proposed techniques are: anisotropic diffusion filter (to remove noise) and subsequent unsharp mask sharpening filter (edge sharpening) [194]. Two-step gamma corrections are used for intensity transformation:

$$f_\gamma(\alpha) = \begin{cases} (c^{\gamma_1-1}\alpha)^{1/\gamma_1} & \text{for } \alpha < c \\ \left(\frac{(2^b-1)^{\gamma_2}-c^{\gamma_2}}{(2^b-1)-c}(\alpha-c)+c^{\gamma_2}\right)^{1/\gamma_2} & \text{for } \alpha \geq c \end{cases}, \quad (4.1)$$

where b is the number of bits (e.g. 16), γ_i are the gamma correction factors and c denotes the density value at which the two transformations interchange.

The image enhancement is followed by segmentation, e.g. by mode method (histogram-based techniques) or more advanced combinations of watershed and active contour methods (edge finding techniques) [194]. If the histogram of absorption values shows a bimodal character, as for the packed bed of calcum carbonate described in chapter 6, the mode method is used for phase segmentation [231, 68]. For material composed of inhomogeneous solid phases, the peaks in the absorption histogram might overlap, as for the packed bed of waste tire particles, and local mode methods might be able to properly segment the CT data.

A (local) threshold value, $\alpha_t(\mathbf{x})$, is determined, corresponding to the absorption value at which the minimum between the two (or multiple) peaks lies.

The mathematical methodology to obtain the continuous representation of the optical density values and consequently a continuous representation of the solid-fluid phase boundary is given in [165]. It basically consists of a 3D linear interpolation of the density values and determination of the region where $\alpha(\mathbf{x}) = \alpha_t(\mathbf{x})$. The gradient of $\alpha(\mathbf{x})$ at the boundary,

$$\hat{\mathbf{n}}_{ji}(\mathbf{x}) = \frac{\nabla \alpha|_\mathbf{x}}{|\nabla \alpha|_\mathbf{x}|}, \quad (4.2)$$

is used to calculate the surface normal. The pore-space indicator function,

$$\Psi(\mathbf{x}) = \begin{cases} 0 & \text{for } \alpha(\mathbf{x}) \geq \alpha_t(\mathbf{x}) \\ 1 & \text{for } \alpha(\mathbf{x}) < \alpha_t(\mathbf{x}) \end{cases}, \quad (4.3)$$

is used to convert the gray level matrix into a 0/1 matrix. Ψ equals one if the point $\mathbf{x}$ lies within the void space and equals zero if it lies within the solid phase.

4.3 Morphological properties

The porous medium, whether naturally formed or fabricated, can be classified based on its solid phase structure. It is consolidated or non-consolidated, and disordered or ordered. The latter is distinguished in anisotropic and isotropic structures, which can be composed of single unit cells or complex random structures [102]. Non-consolidated structures are packed beds (loosely packed) while foams are described as consolidated structures.

4.3. Morphological properties

4.3.1 Numerical determination

Depending on the system and number of involved phases, different properties may be of interest. In the present book the focus lies on two-phase systems. Therefore, porosity, specific surface and size distribution of the single phases and representative elementary volume (REV) are important. REV determines the smallest subvolume of the media, which can be considered as continuum [11]. It is the smallest differentiable volume that results in statistical meaningful local average properties. It is important for calculation purposes and additionally provides information about the statistical relevance, homogeneity and isotropy of the sample. The directional determination of morphological properties gives a general idea of the sample's anisotropy. The concepts of mean intercept length, star length distribution and volume orientation [153, 197, 134] are used for the characterization of morphological anisotropy.

The porosity of the sample is calculated with the aid of statistical (correlation) functions, namely, the two-point correlation function,

$$s_2(r) = \frac{\int_V \int_{4\pi} \Psi(\mathbf{x})\Psi(\mathbf{x}+r\hat{\mathbf{s}}) \mathrm{d}\hat{\mathbf{s}} \mathrm{d}\mathbf{x}}{V 4\pi}, \qquad (4.4)$$

which indicates the probability of two points in a porous sample, separated by a distance r, to be in the void phase [12]. It obeys the following conditions:

$$s_2(r=0) = \varepsilon, \qquad (4.5)$$

$$s_2(r \to \infty) = \varepsilon^2, \qquad (4.6)$$

$$\left.\frac{\mathrm{d}s}{\mathrm{d}r}\right|_{r=0} = -\frac{A_0}{4}, \qquad (4.7)$$

and therefore can be used to determine porosity and specific surface area. The two-point correlation function is calculated by MC sampling. Alternatively, a voxel-based determination of porosity (comparing each voxel with the threshold) and specific surface area can be done. Additionally, the effective porosity and effective specific surface area are introduced to account for the actual accessible volume by the fluid phase, neglecting close and unconnected pores.

REV is determined by calculating morphological, heat and/or mass transfer properties for a number of subsequently growing subvolumes in the sample until the determined properties only vary within a band of $\pm\delta$.

For pore- and particle-size characterization and distribution calculations, the concept of granulometry is applied [134]. It is based on mathematical morphology operations. An opening, which consists of an erosion followed by a dilation with the same structuring element, is used to calculate the pore- and particle-size distribution. For a spherical structuring element with diameter d, the distribution function is determined by

$$f(d) = -\frac{\mathrm{d}\varepsilon_{\mathrm{op}}(d)}{\varepsilon \mathrm{d}d}. \tag{4.8}$$

It determines the size distribution defined by the diameter of a sphere which fits completely in the void and pore space, respectively. To account for anisotropic or non-symmetrical pore or particle shapes nonspherical shaped, directional structuring elements can be used [134]. The hydraulic diameter [102] of the pore and particle is defined as

$$d_{\mathrm{h,pore}} = 4\frac{\varepsilon}{A_0}, \tag{4.9}$$

$$d_{\mathrm{h,particle}} = 4\frac{1-\varepsilon}{A_0}. \tag{4.10}$$

For a more exhaustive morphological characterization additional properties such as local porosity distribution, local specific surface area distribution, local geometry distribution, local percolation probability, total fraction of percolating cells, and chord-length distribution function can be determined [89, 215, 164]. It has been shown that the most promising approaches for the reconstruction of the porous media's morphology are based on combinations of several of the determined morphological properties [179, 128] or based on Minkowski functionals [134].

4.3.2 Experimental determination

Direct and indirect methods are used for porosity measurements. The former methods involve the determination of any of the involved volumes (pore, solid, or total volume) by means of weighing or fluid displacement measurements. Indirect methods involve measurement of some property in the pore space like electrical conductivity of an electrically conducting fluid filling the pore space. Porosity, experimentally determined by weight measurement, is calculated by

$$\varepsilon_{\mathrm{ex}} = 1 - \frac{m/\rho}{V}, \tag{4.11}$$

where m is the measured mass of the sample and V the total volume measured. $\varepsilon_{\mathrm{ex}}$ strongly depends on the density ρ, of the material. The intrinsic and the apparent (including all unconnected pores) density can be determined by helium pycnometry, which is based on fluid displacement measurements in the (partially) crushed material.

4.3. Morphological properties

Adsorption methods are widely used for the experimental determination of the specific surface area, e.g, the Brunauer, Emmit and Teller (BET) method, which determines the BET surface area by nitrogen adsorption [196]. The gas is adsorbed on the solid surface dependent on the pressure imposed. Ideally monolayers of gas are adsorbed and allow to recalculate an equivalent surface area and to estimate pore-size distribution. Additionally, pore-size distribution can be determined by mercury porosimetry [219]. Nevertheless, the interpretation of the data is difficult for non-spherical pores.

Particle-size distributions can be measured by sieving, visual analysis (microscopes, tomography, etc.), sedimentation of particles in liquid, permeametry and laser scattering [177]. The latter approach is used in this study. The particles are suspended in a transparent liquid phase and the scattering behavior is recorded. A sphere-equivalent size distribution is determined based on the refractive indexes of the liquid and the particles, and on the transmitted or scattered light by using the Mie theory [16].

Chapter 5
Reticulate porous ceramics[1,2,3]

Thermochemical and hybrid thermochemical/electrochemical sulfur-based cycles have been introduced as an alternative route for the solar generation of hydrogen at relatively moderate temperatures but in a corrosive environment.

The cycles are introduced in the first part of this chapter. A two-chamber solar reactor for the high-temperature step of the cycles has been designed at DLR consisting of reticulate porous ceramic (RPC) and honeycomb structures, directly exposed to high-flux solar irradiation, serving as solar absorber, heat exchanger and reaction sites for the evaporation, decomposition and subsequent reduction of sulfuric acid and its decomposition products.

The second part of this chapter is devoted to fundamental analysis of the morphological, heat and mass transfer characterization of the RPC used in the evaporator/decomposer reactor. The characterization follows the procedure in [160, 73].

These properties are then incorporated in a coupled continuum model of the solar reactor described in the third part of this chapter. The reactor is analysed

---

[1] Material from this chapter has been published in: S. Haussener, P. Coray, W. Lipiński, P. Wyss, and A. Steinfeld. Tomography-based heat and mass transfer characterization of reticulate porous ceramics for high-temperature processing. *Journal of Heat Transfer*, 132: 023305, 2010. [78]

[2] Material from this chapter has been published in: D. Thomey, M. Roeb, P. Rietbrock, J. Säck, C. Sattler, S. Haussener, A. Steinfeld, I. Canadas and S. Martínez. Development of a two-chamber receiver-reactor for the solar decomposition of sulfuric acid. In *Proceedings of SolarPACES 2009 Conference*, Berlin, 2009. [212]

[3] Material from this chapter has been published in: M. Roeb, Thomey, D. Graf, C. Sattler, S. Poitou, F. Pra, P. Tochon, C. Mansilla, J.-C. Robin, F. Le Naour, R. Allen, R. Elder, I. Atkin, G. Karagiannakis, C. Agrafiotis, A. Konstandopoulos, M. Musella, P. Haehner, A. Giaconia, S. Sau, P. Tarquini, S. Haussener, A. Steinfeld, S. Martinez, I. Canadas, A. Orden, M. Ferrato, J. Hinkley, E. Lahoda and B. Wong. HycycleS - A project on nuclear and solar hydrogen production by sulfur based thermochemical cycles. *International Journal of Nuclear Hydrogen Production and Application*, 2:202–226, 2011. [181]

$$H_2SO_{4,\text{fl}} = H_2SO_{4,\text{g}} = H_2O + SO_3 = H_2O + SO_2 + \tfrac{1}{2}O_2 \quad (5.1)$$
$$2H_2O + SO_2 = H_2SO_4 + H_2 \quad (5.2) \quad \bigg| \quad \begin{aligned} 2H_2O + SO_2 + I_2 &= H_2SO_4 + 2HI \quad (5.3) \\ 2HI &= I_2 + H_2 \quad (5.4) \end{aligned}$$

for chemical and thermal efficiencies and optimized in operation conditions, reactor design and foam morphology.

5.1 Sulfur-based water-splitting cycles

Two sulfur-based water-splitting cycles for the solar production of hydrogen are introduced: a hybrid (thermochemical/electrochemical) sulfur-based cycle and a three-step thermochemical sulfur-iodine based cycle. Both cycles process the same high-temperature step described by the overall chemical reaction in eq. (5.1). Sulfuric acid is evaporated and decomposed at approximately 610 K and the resulting SO_3 is reduced to SO_2 at 1500 K. The temperature of the SO_3 reduction step can be reduced when using catalysts such as Pt, Fe_2O_3, or mixtures of Pt and TiO_2 [65, 149].

In the hybrid sulfur-based cycle [23, 22] the subsequent step is an electrochemical reaction of water and SO_2, described by eq. (5.2). SO_2 is oxidized to H_2SO_4 at the anode ($SO_2 + 2H_2O \rightarrow H_2SO_4 + 2H^- + 2e^-$) and hydrogen is formed at the cathode of the electrolyser ($2H^+ + 2e^- \rightarrow H_2$). Sulfuric acid is re-used in the high-temperature reaction given by eq. (5.1). The net reaction is the splitting of water to oxygen and hydrogen. The electrolyzer requires less than 15% of the electrical power needed for conventional direct electrochemical water splitting (theoretical 0.17 V in reaction (5.2) versus theoretical 1.23 V for conventional direct water splitting).

In the sulfur-iodine based cycle [152, 154, 221, 97], the Bunsen reaction, described by (5.3) and taking place at 400 K, follows the high temperature step, resulting in two immiscible aqueous solutions consisting of aqueous sulfuric acid and HI. They are separated and piped to the decomposition reactions (5.4) and (5.1). The former takes place at temperatures of 400 to 600 K.

The cycles work in a corrosive environment at high temperatures, therefore research has been focused on corrosion and high-temperature resistant materials and membranes for product gas separation. As the temperatures needed for the sulfur-based cycles are moderate compared to other thermochemical water-splitting cycles, the heat source for the high temperature step given by eq. (5.1) proposed initially was waste heat from nuclear power plants. Recently, solar driven evaporation and decomposition have gained attention. Two exemplary research projects funded by the European Commission are mentioned: HYTEC [48] and HycycleS [181].

Equilibrium calculations of the reaction are shown in figure 5.1 for 0.1, 1 and 10 bar. Higher pressure shifts the equilibrium position to the left, due to Le Chatelier's principle.

Energy efficiencies of 47% are achieved for the three-step sulfur-iodine cycle and cost estimates have been done [14] predicting hydrogen prices of 13.5 \$/kg in 1984, strongly dependent on solar irradiation and solar collector field costs. The price is comparable to the price predicted for hydrogen produced by water electrolysis, driven by electricity produced via solar driven Rankine cycles [14]. This price is three to four times higher than the price for hydrogen produced by conventional processes (steam reforming of natural gas).

Solar reactor concepts for the sulfuric acid evaporation and decomposition are based on honeycombs and RPCs [149], and on shell-and-tube configuration with a packed (catalyst) bed [118]. A two chamber reactor, allowing for individual evaporation/decomposition and SO_3 reduction reactions, has been developed in the HycycleS project [212] and is shown in figure 5.2. The evaporation/decomposition reactor uses RPC as radiation absorber and the SO_3 reduction reactor uses a honeycomb structure. The porous structures provide efficient absorption of concentrated solar radiation and large specific surface area for evaporation, decomposition and reduction.

5.2 Characterization of reticulate porous ceramics

5.2.1 Porous ceramics

Porous ceramics are commonly processed by (*i*) template methods or (*ii*) by foaming methods [186, 209]. In the former technique, a polymeric sponge, impregnated with a ceramic slurry (e.g., composed of SiC, Al_2O_3, and binder), is burned out leaving a porous ceramic structure. The remaining structure is sintered at approximately 1400 °C. RPCs are commonly made by this process also known as Schwartzwalder process [192]. They are defined as highly porous artificial structures comprised of interconnected voids surrounded by a web of ceramics. The rehological characteristics of the slurry, namely pseudoplastic and shear-thinning behavior, are crucial for the foam's quality. Recently monofilamental pre-structures are investigated, which allow for a larger variety and better controllability of the foam structures [6]. Since open cell foams produced by the Schwartzwalder process show hollow spaces in their struts, as shown in

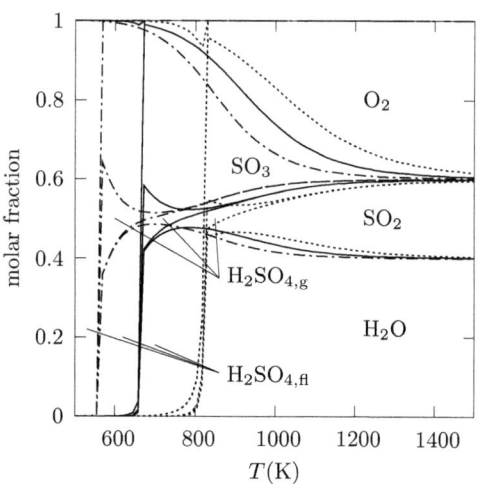

Figure 5.1: Equilibrium composition of sulfuric acid evaporation and decomposition at 0.1 bar (dash dotted line), 1 bar (solid line), and 10 bar (dotted line).

figure 5.3.a, and consequently reduced mechanical strength and stability, techniques are proposed [7] to fill this hollow spaces, see figure 5.3.b.

The template method is also used to produce ceramic material of the negative replica of the original template, as opposed to the positive morphology obtained by the method described above. This is obtained by the preparation of biphasic composites comprising a continuous matrix of ceramic particles and a discontinuous sacrificial phase, which is ultimately extracted to generate pores. The anisotropic porous ceria foam, described in chapter 6, is manufactured by this process.

The foaming technique to process porous ceramics combines reactants, including the desired ceramic constituents, to produce a foam from the evolved gas. This direct foaming method can produce open and closed-cell foams. Examples of porous ceramics produced by the template and the foaming method are shown in figure 5.4.

Porous ceramics show advantageous performance in high-temperature, extensive wear and corrosive environments. Some of the favourable morphological, heat and mass transfer characteristics of porous ceramics are: relatively low mass, low density, high specific surface area, high melting point, high corrosion and wear resistance, low thermal mass, low thermal conductivity, and controlled permeability. These properties can be additionally tailored for each

5.2. Characterization of reticulate porous ceramics

(a)

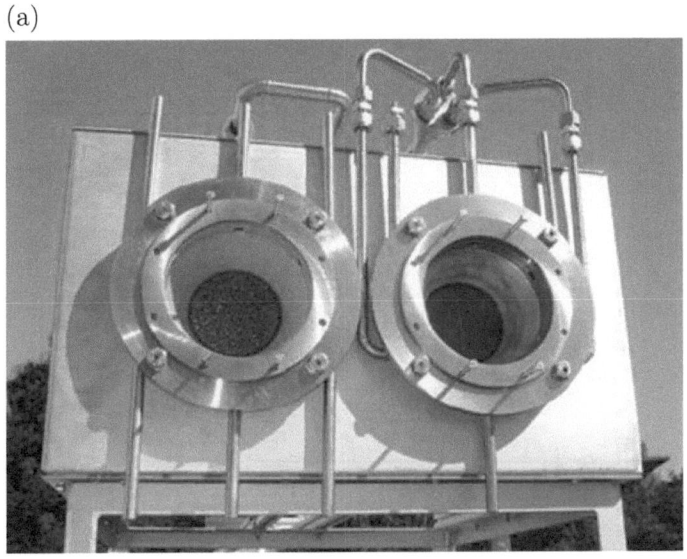

(b)

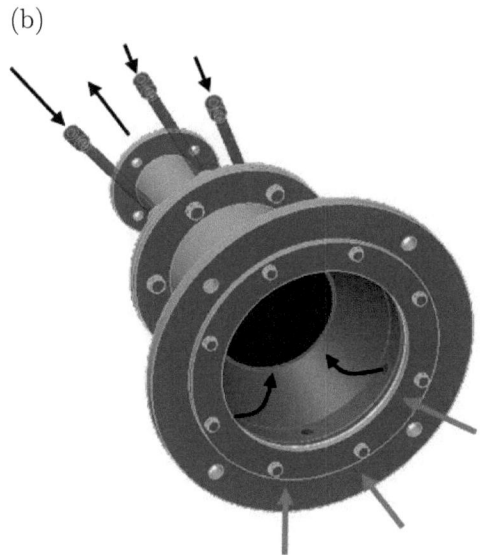

Figure 5.2: (a) Photo of the two-chamber evaporation/decomposition (left side) and SO_3 reduction reactor (right side), and (b) schematic of the evaporation chamber hosting RPC [212].

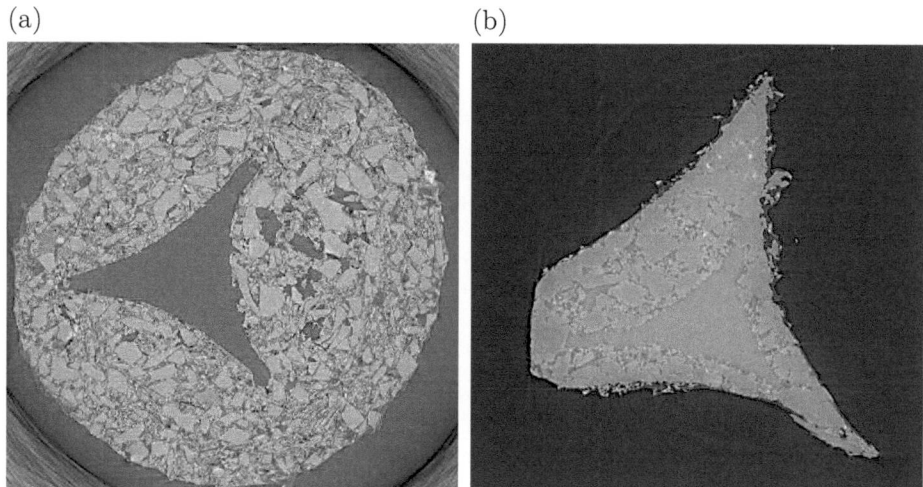

Figure 5.3: Tomographic image of RPC samples: Strut of SiC RPC produced by conventional Schwartzewalder process, edge length 711 µm (a), and strut of SiSiC RPC produced by LigaFill process, edge length of 281 µm (b).

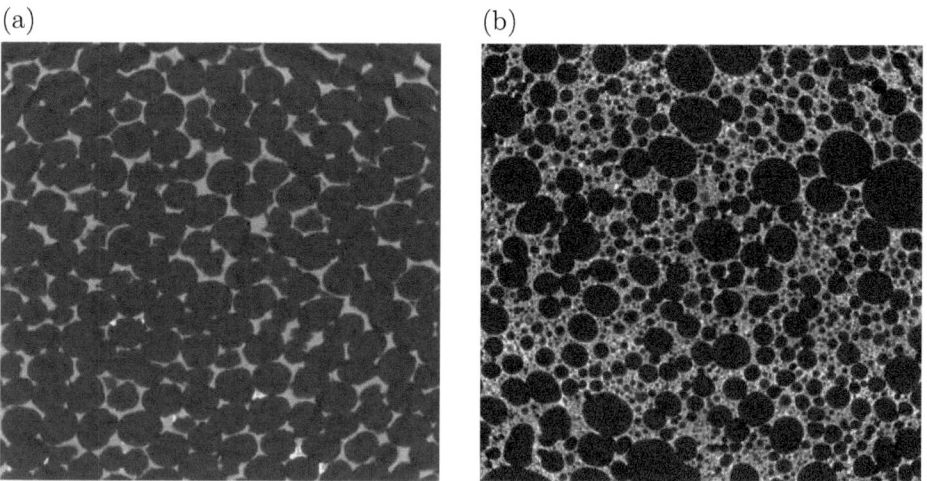

Figure 5.4: Tomographic image of an open-cell SiSiC RPC (a) and partially closed-cell Al_2O_3 foam (b). Edge length of the pictures are 2.3 cm.

5.2. Characterization of reticulate porous ceramics

specific application by controlling the composition and morphology of the porous ceramic and consequently by controlling its processing route. Applications of porous ceramics are molten-metal and diesel-engine-exhaust filters [132, 222], radiant burners [10], catalyst support [46], filtration of hot corrosive gases, high-temperature thermal insulation and radiant absorbers in solar thermal and thermochemical reactors [53].

Previous pertinent studies on heat and mass transfer characterization of foams include the determination of the extinction coefficient, scattering coefficient, and scattering phase function of metal foams and RPCs for simplified geometries composed of pentagon dodecahedron or tetracaedecaedric [122] and spherical voided cubic unit cells [59]. The effective thermal conductivity has been determined for simplified foam geometries composed of tetrakaidecahedron [18], hexagonal [13], and spherical or cubical voided cubic unit cells [60]. Permeability has been determined by simplifying the foam structure by parallel conduits [102] or by empirical correlations for fibrous beds [49, 29]. The exact geometry of a porous magnetic gel sample, obtained by micro-CT, was used to compute the mean survival time, which was linked to the permeability by an empirical correlation [178]. Residence time and tortuosity distribution of real (obtained by magnetic resonance imaging) and artificially generated open-cell foams have been determined in [73]. More recently, CT-based methodology was applied to determine the effective transport properties of a Rh-coated SiC RPC used for the solar steam-reforming of CH_4 [165, 164, 163, 162].

5.2.2 Morphological characterization

An uncoated, non-hollow SiSiC RPC with nominal pore diameter of $d_{\text{nom}} = 1.27$ mm (corresponding to 20 pores per inch, 20 ppi) is used in the solar reactor. SiSiC is produced by doping SiC with C and infiltrating Si. The Si and C produce a matrix of SiC and its pores are filled with the remaining Si. The RPC foam sample is shown in figure 5.5.a and used for the subsequent analysis.

Low-resolution computer tomography

The sample is exposed to an unfiltered polychromatic X-ray beam generated by electrons incident on a wolfram target. The generator is operated at an acceleration voltage of 60 keV and a current of 0.11 mA. A Hamamatsu flatpanel C7942 CA-02 protected by a 0.1 mm thick aluminum filter is used to detect the transmitted X-rays. The sample is scanned at 900 angles (projections). Each projection consists of an average of six scans, each with an exposure time of

0.7 s. Low-resolution computer tomography (LRCT) is performed for voxel sizes of 30 µm (shown in figure 5.4.a) and 15 µm and field of view (FOV) of 2.3 x 2.3 x 1.2 cm^3 and 1.2 x 1.2 x 0.6 cm^3, respectively. 3D surface rendering of a sample subvolume, reconstructed from the resulting tomography data, is depicted in figure 5.5.b.

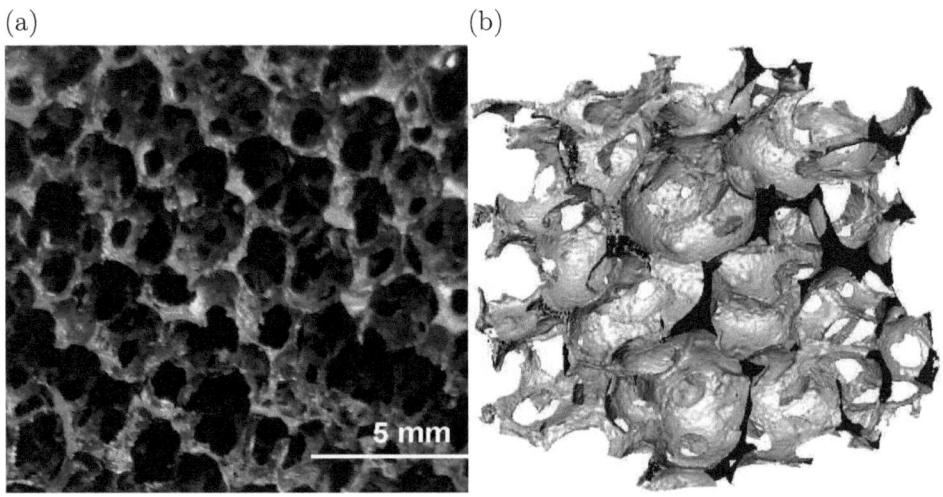

Figure 5.5: Photograph of RPC sample (a), and 3D surface rendering of the RPC sample obtained by using the CT data with voxel size of 15 µm, the edge length is 3 mm (b).

The histograms of the normalized absorption values, as shown in figure 5.6.a, have a bimodal character for both scan resolutions. The mode method is applied for phase segmentation [231, 68]. The normalized phase threshold values $\alpha_t/\alpha_{\max}$ are 0.39 and 0.23 for scans with voxel sizes of 30 µm and 15 µm, respectively. These results are comparable to $\alpha_t/\alpha_{\max} = 0.35$ and 0.20 for scans with voxel sizes of 30 µm and 15 µm, respectively, calculated using Otsu's method [68]. $\alpha_t/\alpha_{\max}$ varies by 0.04, as corroborated for three selected tomograms divided into 36 subelements and for voxel sizes of 15 µm and 30 µm, shown in figure 5.6.b.

High-resolution computer tomography

High-resolution computer tomography (HRCT) of a single RPC strut with 0.37 µm voxel size and 0.76 x 0.76 x 0.62 mm^3 FOV is shown in figure 5.3.b. The

5.2. Characterization of reticulate porous ceramics

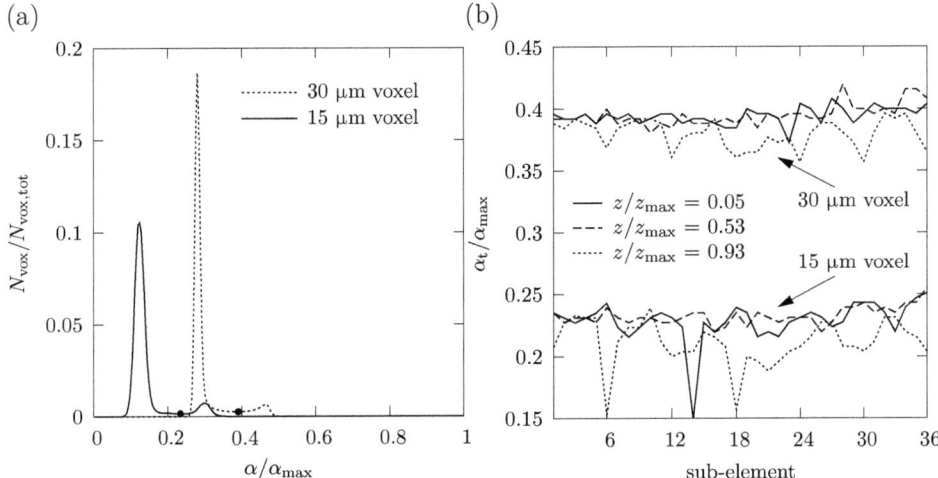

Figure 5.6: (a) Normalized histograms of the absorption values for the HR and LR scans. The bullets indicate the corresponding threshold values of $\alpha_t/\alpha_{max} = 0.39$ and 0.23. (b) Normalized threshold absorption values for 36 subelements of three selected tomograms for voxel sizes of 30 µm and 15 µm.

HRCT is performed on the TOMCAT beamline at the SLS of the PSI [208, 200] for 14 keV photon energy, 400 µA beam current, 100 µm thick aluminum filter, 20 x geometrical magnification, 0.8 s exposure time, and 1501 projections. A magnified fragment of a strut edge is shown in figure 5.7. The strut surface is irregularly shaped. Irregular spatial distribution of SiC and Si within the strut leads to internal heterogeneity, but no pores are observed inside the strut.

Porosity and specific surface

The two-point correlation function is computed by MC sampling with 10^8 random points and for r varying between 0 and 1 cm. The resulting porosity is $\varepsilon = 0.91$. It compares well to the value obtained by weight measurement (0.90 ± 0.02) and it does not depend on the voxel size of the scan. In contrast, the specific surface area A_0 increases from 1367 m^{-1} to 1680 m^{-1} as the voxel size decreases from 30 µm to 15 µm because of the better resolution of surface irregularities for the smaller voxel size.

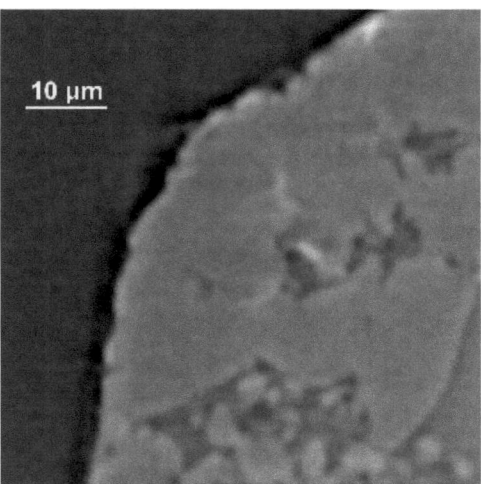

Figure 5.7: Magnified fragment of the strut edge.

Representative elementary volume

REV is defined as the minimum volume of a porous material for which the continuum assumption is valid. It is determined based on porosity calculations for subsequent growing volumes until it asymptotically reaches a constant value within a band of $\pm\delta$. For $\delta = 0.02$, $l_{\mathrm{REV}} = 3.50$ mm and 3.39 mm for the scans with voxel sizes 30 µm and 15 µm, respectively, corresponding to $2.76 d_{\mathrm{nom}}$ and $2.67 d_{\mathrm{nom}}$, respectively. It will be shown in the analysis in section 5.2.3 that at least $l_{\mathrm{REV}} = 3.50 d_{\mathrm{nom}}$ is required for computations of heat and mass transfer.

Pore size distribution

An opening operation with spherical structuring element (diameter d) is applied to compute the opening porosity $\varepsilon_{\mathrm{op}}$ as a function of d and then used to determine the RPC's pore size distribution function f. The resulting pore size distribution functions are shown in figure 5.8 for the 30 µm and 15 µm voxel size tomography data. Table 5.1 lists the mean, mode, median, and hydraulic diameters resulting from the distributions in figure 5.8. Good qualitative and quantitative agreement of porosity, specific surface area, and pore size distribution is obtained for voxel sizes of 30 and 15 µm. Thus, the heat and mass transfer characteristics computed in the following sections are done by using the 30 µm resolution tomography data.

5.2. Characterization of reticulate porous ceramics

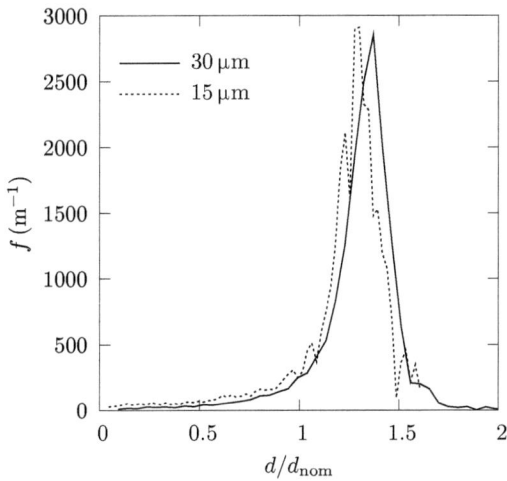

Figure 5.8: Opening pore size distribution of the RPC foam for the LR and HR tomography data.

Table 5.1: Arithmetic mean diameter, mode, median, and hydraulic diameter for the 30 µm and 15 µm voxel size tomography data.

Voxel size (µm)	d_m (mm)	d_mode (mm)	d_median (mm)	d_h (mm)
30	1.64	1.74	1.69	2.76
15	1.55	1.65	1.62	2.24

5.2.3 Heat transfer characterization

Radiative properties

The complex refractive index of SiC is used for the calculation of solid phase's intrinsic radiative properties [158]. SiC is assumed to be opaque in the visible and near-IR spectral region. The fluid phase is assumed to be radiatively non-participating. Hence, the governing equations and the methodology for the determination of the radiative properties described in chapters 2 and 3 are applied. $G_\text{e}(s)$ and $F_{\mu_\text{in}}(\mu_\text{in})$ are computed by following the histories of a large number of stochastic rays ($N_\text{ray} = 6 \cdot 10^6$) launched at random locations within the fluid phase of the REV. N_e rays interact with the solid-fluid interface by either absorption or reflection. The path length s within REV is recorded for all N_ray rays, while the cosine of incidence μ_in is recorded for all N_e rays [211, 165].

Table 5.2: Mean values and standard deviations of the extinction coefficient along three directions.

	x-direction	y-direction	z-direction
β_{MC}, mean (m^{-1})	400.6	411.0	439.8
$S_{\beta,i}$ (m^{-1})	35.8	37.5	28.4

A sample of 18 x 18 x 12 mm^3, corresponding to 600 x 600 x 400 voxels, is investigated. $1 - G_{\mathrm{e}}(s)$ is plotted in figure 5.9.a as a function of the normalized path length for two values of the threshold $\alpha_{\mathrm{t}}/\alpha_{\mathrm{max}}$: 0.31 and 0.39. A least-square fit to Bouguer's law, also shown in figure 5.9.a, yields a constant extinction coefficient $\beta_{\mathrm{MC}} = 628.4$ m^{-1} with RMS = 0.014 m^{-1} for $\alpha_{\mathrm{t}}/\alpha_{\mathrm{max}} = 0.31$, and $\beta_{\mathrm{MC}} = 430.8$ m^{-1} with RMS = 0.014 m^{-1} for $\alpha_{\mathrm{t}}/\alpha_{\mathrm{max}} = 0.39$. Separate computations along preferred directions showed slight anisotropy of β, as indicated in table 5.2.

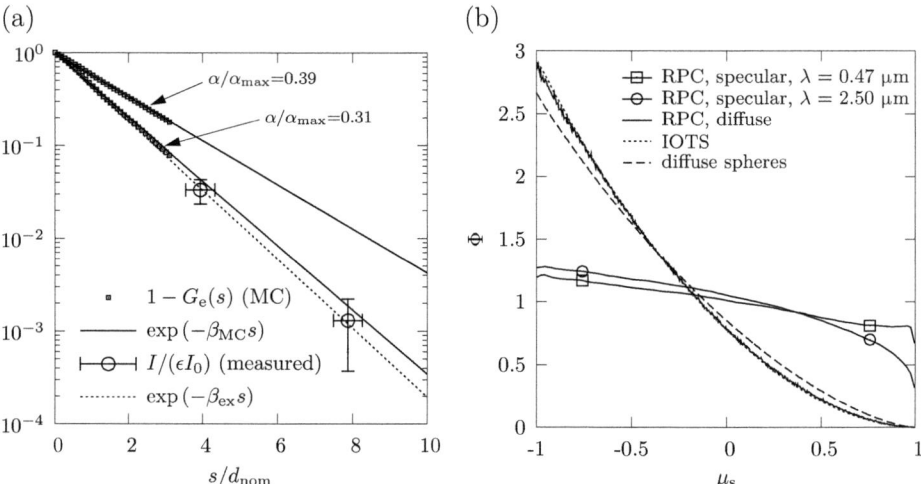

Figure 5.9: (a) Variation in computed and measured incident radiative intensities as a function of normalized path length in the sample; (b) scattering phase functions of the RPC foam, IOTS, and of large diffuse opaque spheres as a function of the cosine of scattering angle.

For assumed diffusely reflecting surface of the solid phase, the scattering phase function is plotted in figure 5.9.b as a function of the cosine of scattering

5.2. Characterization of reticulate porous ceramics

angle. It is well approximated by (RMS = 0.010)

$$\Phi_d = 0.5471\mu_s^2 - 1.38838\mu_s + 0.8176. \tag{5.5}$$

Also shown in figure 5.9.b is the analytically determined scattering phase function for diffusely reflecting large opaque spheres [141]. The RPC foam and identical overlapping transparent spheres (IOTS) [211] exhibit identical scattering behavior due to their morphological similarity. Compared with large diffuse opaque spheres, RPC exhibits enhanced scattering in backward direction and less in forward direction. For assumed specularly reflecting surface, two exemplary scattering phase functions computed for $m_{\lambda=0.48\mu m} = 2.707 + 1.46i$ and $m_{\lambda=2.50\mu m} = 2.562 + 5.17i$ are shown in figure 5.9.b. Both exhibit nearly isotropic scattering behavior with slightly increased backward scattering. Based on the irregular surface topography shown in figure 5.7, the solid phase is expected to show a dominant diffuse component in the reflection pattern.

The scattering albedo σ_s/β equals the surface reflectivity of the solid phase, assumed to be wavelength independent and equal to 0.1 [216]. Hence, $\sigma_s = 43.1$ m^{-1} and 62.8 m^{-1} for $\alpha_t/\alpha_{max} = 0.31$ and 0.39, respectively.

The extinction coefficient is independently estimated based on experimental measurements by using the spectroscopy system shown in figure 5.10 [39]. The main hardware components of the setup are as follows: (1) a dual Xe-Arc/Cesiwid-Glowbar lamp as a source of radiation, (2) a double monochromator (Acton Research Spectra Pro Monochromator SP-2355 series) with monochromator exit slit (2'), (3) and (5) two imaging lens pairs (MgF$_2$, focal lengths f = 75 mm and f = 150 mm), (4) a sample holder, (6) a detector (Si/PC-HgCdTe sandwich with thermoelectric cooler), (7) an optical chopper to modulate the radiation leaving the monochromator, (8) a lock-in amplifier to measure the modulated signal, and (9) a PC data acquisition system. This setup enables measurements in the spectral range from 0.3 µm to 4 µm with a spectral resolution of ± 1 nm and an angular resolution of 10°. The maximum acceptance angle for detection of an incoming ray measured with respect to the optical axis is less than 4°. Angular measurements are performed with two RPC foam samples of thicknesses 5 mm and 10 mm at radiation wavelengths of 0.3 µm, 0.6 µm, and 0.9 µm. The measured flux rapidly decreases with the increasing detection angle. At 10°, it is 10^3 smaller than that acquired in the forward direction. Since the reflectivity of the solid phase is comparable to that of SiC (0.1) [216], the contribution of the incoming scattering to the measured radiative fluxes is neglected. Thus, the extinction coefficient is estimated by assuming Bouguer's law-type dependency of the measured radiative fluxes on the sample thickness.

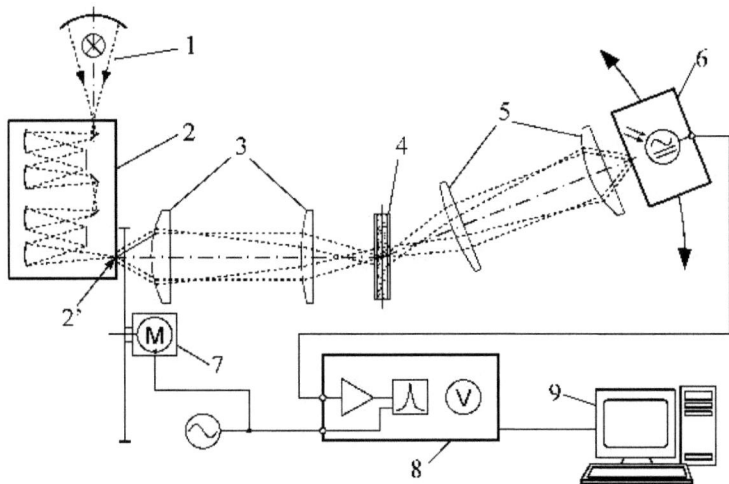

Figure 5.10: Experimental spectroscopy setup: (1) dual Xe-Arc/Cesiwid-Glowbar lamp, (2) double monochromator, (3) and (5) collimating and focusing lens pairs, (4) sample mounted on alinear positioning stage, (6) detector, (7) optical chopper, (8) lock-in amplifier, (9) data acquisition system.

For all radiation wavelengths, the extinction coefficient is approximately constant and equal to $\beta_{ex} = 673 \pm 30$ m^{-1}. This value is larger than $\beta_{MC} = 430.8$ m^{-1} determined numerically by MC for $\alpha_t/\alpha_{max} = 0.39$, but it is in good agreement with $\beta_{MC} = 628.4$ m^{-1} computed for $\alpha_t/\alpha_{max} = 0.31$. Since $1/\sigma_s$ has the same order of magnitude as the measured sample thickness, incoming scattering may affect the experimental results. In contrast, when comparing measured and calculated porosities, $(\varepsilon_{ex} - \varepsilon_{MC})/\varepsilon_{ex} = 0.01$ and 0.06 for $\alpha_t/\alpha_{max} = 0.39$ and 0.31, respectively.

Effective thermal conductivity

The governing equations and the methodology for the determination of the effective thermal conductivity are described in chapters 2 and 3. A sample of 10.8 x 10.8 x 10.8 mm^3, corresponding to 360 x 360 x 360 voxels, is investigated. Grid convergence is obtained with a mesh element size of 21.5 μm. A contour map of the normalized temperature distribution along the axis perpendicular to the temperature boundary condition is shown in figure 5.11.a. The effective sample conductivity as a function of the solid and fluid conductivity is shown in figure 5.11.b. It is compared with the parallel and serial slab assumptions (at ε

5.2. Characterization of reticulate porous ceramics

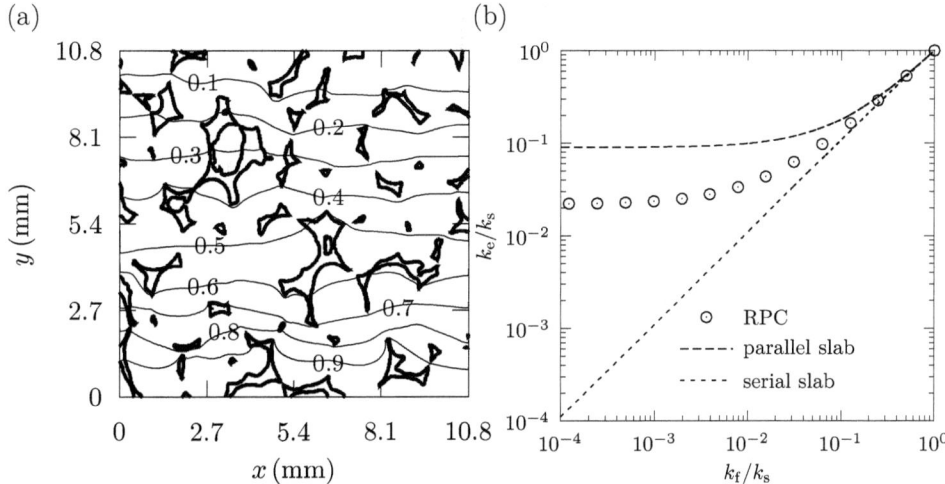

Figure 5.11: (a) Contour map of the normalized temperature distribution $(T - T_2)/(T_1 - T_2)$ along the axis perpendicular to the temperature boundary condition of the RPC foam (thick solid lines depict solid-fluid phase boundary) for $k_f/k_s = 1.0 \cdot 10^{-4}$; (b) the effective thermal conductivity of the RPC foam and of parallel and serial slabs at $\varepsilon = 0.91$.

$= 0.91$) [102], which indicate minimal and maximal possible conductivities. The computed k_e/k_s was fitted to the linear combination of thermal conductivities of parallel and serial slabs,

$$\frac{k_e}{k_s} = a_1 \frac{\frac{k_f}{k_s}}{\varepsilon\left(1 - \frac{k_f}{k_s}\right) + \frac{k_f}{k_s}} + a_2 \left(\varepsilon \frac{k_f}{k_s} + 1 - \varepsilon\right), \tag{5.6}$$

resulting in $a_1 = 0.753$ and $a_2 = 0.267$. a_1 and a_2 depend strongly on the morphology. This can be seen when comparing k_e of this 20 ppi and the 10 ppi foams analyzed in [165], both having the same ε and a rather sharp pore size peak (peak width $\approx 0.5 d_{\text{nom}}$). The 20 ppi foam shows a larger A_0. For $k_f/k_s \approx 10^{-1}$ they differ by nearly 6% and at $k_f/k_s \approx 10^{-4}$ they differ by up to 40%. Obviously, the smaller k_f/k_s is, the more important becomes the phase distribution. When using the RPC foam as solar absorber, larger k_e are preferred because they allow for a faster heat transfer rate and a more uniform heating.

Interfacial heat transfer coefficient

The heat flux from the solid to the fluid phase is described by the heat transfer coefficient. The governing equations and the methodology for the determination of the interfacial heat transfer coefficient are described in chapters 2 and 3. A sample with dimensions of 11.4 x 11.4 x 3.78 mm^3, corresponding to 380 x 380 x 126 voxels, is investigated. Convergence was achieved for the termination residual RMS of the iterative solution below 10^{-4} and for the maximal mesh element length of 90 µm (corresponding to $0.07 d_{\text{nom}}$), resulting in $5.6 \cdot 10^7$ tetrahedral elements. Two quad-core Intel Xeon 2.5 GHz processors and 32 Gbytes RAM are used to solve the equations in approximately 10 h. Nu is shown in

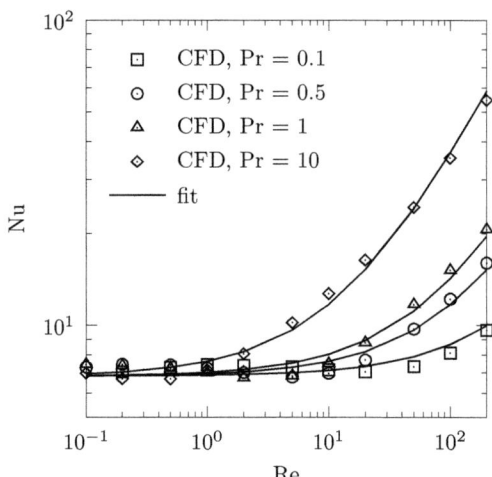

Figure 5.12: Computed (points) and fitted (lines) Nu numbers as a function of Re and Pr numbers.

figure 5.12 as a function of Pr and Re. Assuming a correlation of the form $\text{Nu} = a_1 + a_2 \text{Re}^{a_3} \text{Pr}^{a_4}$, least-square fitting leads to (RMS = 0.817):

$$\text{Nu} = 6.820 + 0.198 \text{Re}^{0.788} \text{Pr}^{0.606}. \tag{5.7}$$

These results compare well to those obtained experimentally for a 10 ppi foam [25].

Influence of l_{REV}

Normalized porosity, extinction coefficient, and conductivity are plotted in figure 5.13 as a function of l/d_{nom} varying between 0 and 6. All three parameters

5.2. Characterization of reticulate porous ceramics

converge for approximately $l_{REV} = 3.5d_{nom}$, with $\delta = \pm\ 0.02, \pm\ 0.02$, and $\pm\ 0.15$, respectively.

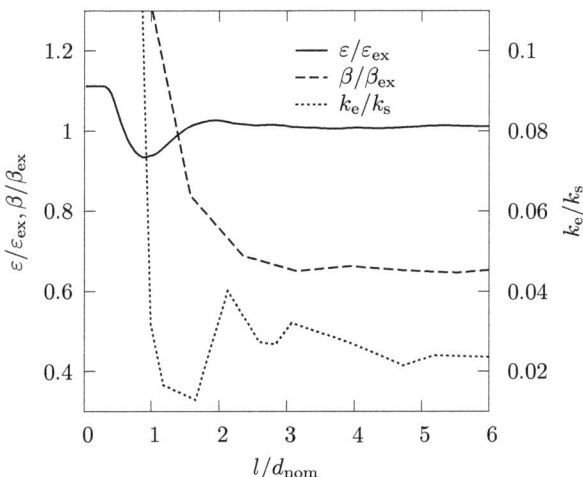

Figure 5.13: Normalized porosity, extinction coefficient, and effective conductivity for cubic volumes with edge lengths l.

5.2.4 Mass transfer characterization

Permeability and Dupuit-Forchheimer coefficient

The momentum transfer within a porous media is described by permeability K, and Dupuit-Forchheimer coefficient F_{DF}. The governing equations and the methodology for the determination of K and F_{DF} are described in chapters 2 and 3. A 11.4 x 11.4 x 7.56 mm³ sample, corresponding to 380 x 380 x 252 voxels, is investigated. Convergence is achieved for the termination residual RMS of the iterative solution below 10^{-4} and for the maximal mesh element length of 180 µm (corresponding to $0.14d_{nom}$), resulting in $5.6 \cdot 10^7$ tetrahedral elements.

The dimensionless pressure gradient is plotted as a function of Re in figure 5.14. Least-square fitting (RMS = 0.566) results in $a_1 = 28.334$ and $a_2 = 0.659$ of eq. (3.24), which correspond to $K = 5.69 \cdot 10^{-8}$ m² and $F_{DF} = 519.0$ m⁻¹. Permeability values are compared with those obtained for the Hagen-Poisseuille model ($4.59 \cdot 10^{-8}$ m², see eq. (7.25)) [102], the Carman-Konzeny model ($8.69 \cdot 10^{-8}$ m², see eq. (7.26)) [49, 102], the fibrous bed models ($8.97 \cdot 10^{-7}$ m², see eq. (7.27)) [102], and a model proposed by Macdonald et al.

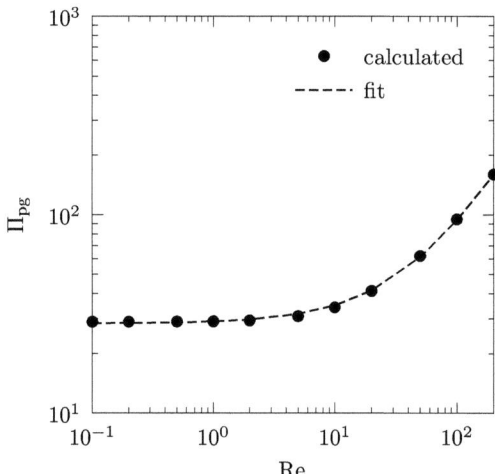

Figure 5.14: Dimensionless pressure gradient as a function of Re number.

$(5.34 \cdot 10^{-8}$ m$^2)$ [126]. The latter model yields $F_{\text{DF}} = 544.2$ m^{-1}, see eq. (7.31). The inertia induced term comes into play for Re > 1. Increased pore size leads to reduced pressure loss and, consequently, larger K and smaller F_{DF}. This can be observed when comparing this 20 ppi SiSiC RPC to the 10 ppi Rh-coated SiC analyzed in [162].

Tortuosity and residence time distributions

The velocity distributions obtained by discrete-scale numerical calculations are used to determine the tortuosity and residence time distributions. They are shown in figure 5.15 for Re = 0.1, 1, 10, and 100. Mean tortuosity is 1.07. The peak of the tortuosity distribution shifts toward $\tau = 1$ when Re is increased. Mean residence time is plotted in figure 5.16.a.

All stream lines generated are able to pass through the sample, as no recirculation zones, dead ends, or flow reversals are observed. In most chemical applications, large τ, smaller Re, and consequently large t are preferred to allow for complete reaction conversion.

Dispersion tensor

The governing equations and the methodology for the determination of the dispersion tensor are described in chapters 2 and 3. D_x, D_y, and D_z are determined by following 2500 streamlines through the foam, registering their spatial

5.3. Continuum model

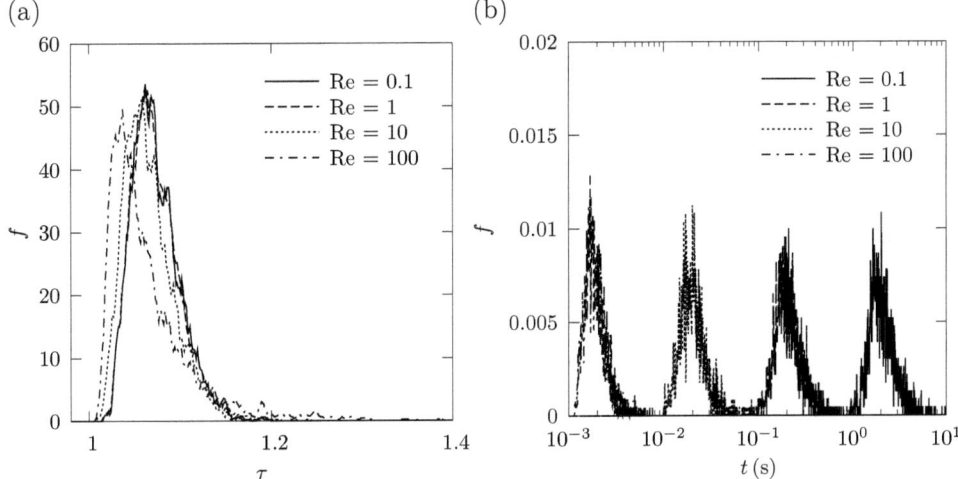

Figure 5.15: (a) Tortuosity and (b) residence time distribution for four selected Re numbers of fluid flow through the RPC foam.

displacement at a specific time instant, and subsequently fitting the registered distribution to a standard Gauss distribution. The normalized dispersion tensor components are shown in figure 5.16.b as a function of Re. D_x and D_y are equal to the transverse (radial) component $D_\perp$; D_z is equal to the parallel (axial) component $D_\parallel$. Fitting to a form $D_i = a_1 \mathrm{Re}^{a_2}$ in two Re regions yields:

$$\frac{D_\perp}{\nu} = \begin{cases} 6.560 \cdot 10^{-3} \mathrm{Re} & \mathrm{Re} \leq 5 \\ 4.896 \cdot 10^{-3} \mathrm{Re}^{1.104} & \mathrm{Re} > 5 \end{cases} (\mathrm{RMS} = 6.0 \cdot 10^{-6}), \qquad (5.8)$$

$$\frac{D_\parallel}{\nu} = \begin{cases} 6.297 \cdot 10^{-1} \mathrm{Re} & \mathrm{Re} \leq 5 \\ 7.045 \cdot 10^{-1} \mathrm{Re}^{0.942} & \mathrm{Re} > 5 \end{cases} (\mathrm{RMS} = 4.2 \cdot 10^{-5}). \qquad (5.9)$$

5.3 Continuum model[4]

A continuum model of the solar evaporator/decomposer reactor, introduced in the first part of this chapter, is developed in this section. The previously calculated morphological and effective heat and mass transfer properties are incorporated in the governing equations. The model is compared to experimental data,

---
[4]Material from this chapter is partially based on work performed in the framework of: J. Marti, Modeling of a solar evaporator reactor for the high-temperature step of a thermochemical cycle for the production of hydrogen, Semester thesis, ETH Zurich, 2010. [131]

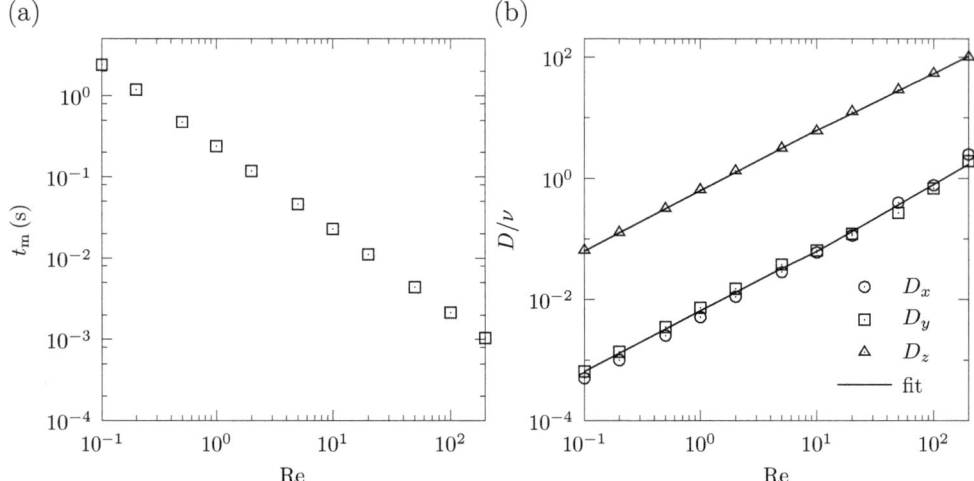

Figure 5.16: (a) Mean residence time as a function of Re number. (b) Normalized dispersion tensor as a function of Re for the RPC foam.

which has been obtained at the solar furnace of Cologne at DLR [212]. Process optimization is carried out by parameter studies on operational conditions, reactor design and foam morphology.

5.3.1 Model development

Two models are investigated: a foam model and a reactor model. The former is used for preliminary studies and for the comparison with experimental data. It is then extended by the reactor closing and further used for the process optimization. The 3D model domains of the foam model and the reactor model are shown in figure 5.17.

The solar reactor consists of a quartz window through which the solar irradiation enters the reactor. A metallic body consisting of a conical front section and a cylindrical back section shapes the reactor. The conical front section is covered by a gold coating to reduce corrosion-related damage. Six radial carrier gas (N_2) inlets behind the window reduce the product deposition at the window. The foam, mounted in the conical part of the metallic body, is directly irradiated by the solar radiation and acts as absorber, heat exchanger, and reaction site. Three tubes enter the foam and deliver the acid solution to evaporate. The carrier gas and the products leave the reactor at the backside. The reactor is covered by an alumina insulation to reduce heat losses to the surrounding.

5.3. Continuum model

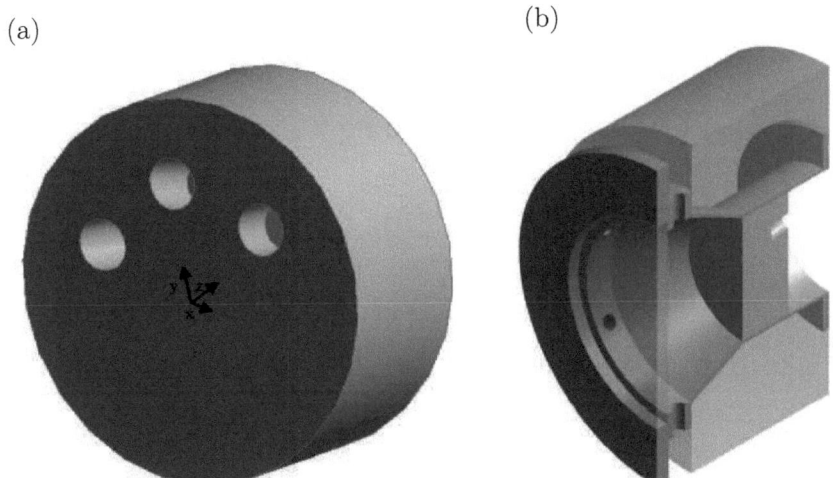

Figure 5.17: 3D model domain of foam model with foam diameter of 9 cm (a), and of reactor model with outer diameter of 25 cm (b).

Averaging models are used for the foam, therefore the foam is modeled as two interpenetrating phases: a void phase and a solid phase. Within the foam, the averaged mass conservation, species conservation, momentum conservation, and energy conservation are valid. Additionally, for the reactor model the void space between the window and the foam, the aperture, the window, the metallic shields, and the insulation have to be modeled. These are single phase domains, therefore they are described by the mass, species, momentum, and energy conservation for one phase. Conservation equations valid for the two-phase foam and the remaining single phase media are given in sections 2.2 and 2.3. FV technique is used to solve the governing equations [92]. Discrete ordinate (DO) method is used to solve RTE [141]. A dual-cell approach is used for the two-phase coupling [121].

The boundary conditions applied in the foam model are the temperature, mass flow, and concentration at the three acid inlets,

$$\mathbf{u} \cdot \hat{\mathbf{n}} = -u_{\text{acid,in}}, T_\text{f} = T_{\text{acid,in}}, x_{\text{acid}} = 1, \qquad (5.10)$$

the mass flow, concentration and temperature of the carrier gas, and the solar irradiation at the front of the foam (xy-plane at $z = l$),

$$\mathbf{u} \cdot \hat{\mathbf{n}} = -u_{\text{carrier,in}}, T_\text{f} = T_{\text{carrier,in}}, x_{\text{acid}} = 0, \mathbf{q}'' = -q''_{\text{rad,in}}, \qquad (5.11)$$

the outlet pressure at the foam outlet,

$$p = p_{\text{atm}}, \qquad (5.12)$$

and no-slip boundary and heat transfer coefficient accounting for conduction and natural convection losses [93] at the outer lateral walls,

$$\mathbf{u} \cdot \hat{\mathbf{n}} = 0, \hat{\mathbf{n}} \cdot \nabla \mathbf{u} = 0, \mathbf{q}'' = q''_{\text{cond}} + q''_{\text{nc}}. \qquad (5.13)$$

At the inlet and outlet the surrounding is assumed to be radiative non-participating. For simplification, the reactor walls are assumed to be perfect reflectors, as justified by the high reflectivity of alumina [216]. The spatial flux distribution at the inlet is obtained by fitting a Gaussian distribution to the experimentally measured flux distribution in the solar furnace, shown in figure 5.18. The di-

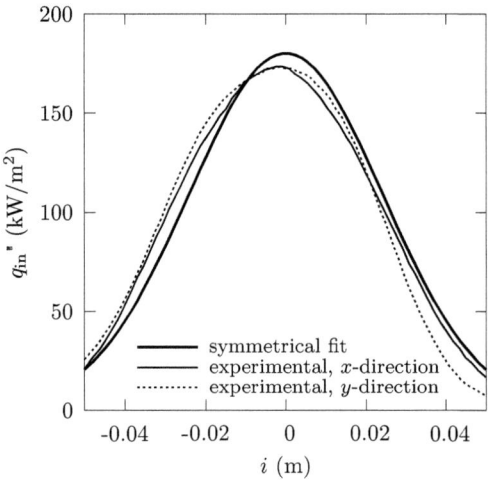

Figure 5.18: Meassured spatial flux distribution and Gaussian fit for the measurements with a target of 0.04 m² and 0.55 kW total power (RMS = 5.4).

rectional distribution of the incident radiation of solar furnaces composed of parabolic dish concentrators can be approximated by a cone-shaped distribution, as is supported by modeling and measurements [111, 240]. For the DLR solar furnace, the directional distribution is assumed to be uniform in a cone with half opening angle of 25°. The flux incident is reduced by 8% to account for the window transmittance in the wavelength region of interest,

$$Tr_\lambda = \tau_\lambda \frac{1 - \rho'^\cap_{r,\lambda}}{1 + \rho'^\cap_{r,\lambda}} \frac{1 - \rho'^{\cap\,2}_{r,\lambda}}{1 - \rho'^{\cap\,2}_{r,\lambda}\tau_\lambda^2}, \qquad (5.14)$$

5.3. Continuum model

$$\rho_{r,\lambda}'^{\cap} = 0.5\frac{\sin^2(\theta - \chi)}{\sin^2(\theta + \chi)}\left(1 - \frac{\cos^2(\theta + \chi)}{\cos^2(\theta - \chi)}\right), \text{ with } \frac{\sin \chi}{\sin \theta} = \frac{n_{1,\lambda}}{n_{2,\lambda}}, \quad (5.15)$$

$$\tau_\lambda = \exp\left(-\kappa_\lambda \frac{l_{\text{sample}}}{\cos \chi}\right). \quad (5.16)$$

The overall transmitted fraction Tr, is calculated by the surface reflectivity ρ_r, and transmittance τ of the window. Each is a function of the wavelength of the incident radiation λ, and the refractive index of air and glass n_1 and n_2, respectively. θ and χ describe the inlet and outlet angle of the ray with respect to the surface normal vector.

The incoming sulphuric acid flow is diluted with water. A water weight fraction of 50% is used in the calculations. The diluted acid is modeled as a mixture of water and sulphuric acid and its properties are calculated by mass-weighted summation of the specific properties. It is assumed that the diluted sulphuric acid entering the reactor is instantaneously evaporated and decomposed. Therefore, a two-phase medium model is used. The evaporation is modeled by a heat sink of the form,

$$q_{\text{sink}} = \dot{m}_{\text{acid}}\left(\sum_i x_i\left(h_{i,\text{l},T_{\text{evap}}} - h_{i,\text{l},T_{\text{in}}}\right) + \Delta h_{\text{evap}} + \Delta h_{\text{decomp}}\right), i = H_2SO_4, H_2O, \quad (5.17)$$

where

$$\Delta h_{\text{evap}} = \sum_i x_i\left(h_{i,\text{g},T_{\text{evap}}} - h_{i,\text{l},T_{\text{evap}}}\right), i = H_2SO_4, H_2O, \quad (5.18)$$

$$\Delta h_{\text{decomp}} = x_{H_2SO_4}\sum_i \frac{M_i}{M_{H_2SO_4}} h_{i,\text{g},T_{\text{evap}}} - h_{H_2SO_4,\text{g},T_{\text{evap}}}, i = SO_3, H_2O, \quad (5.19)$$

and $\dot{m}_{\text{acid}}$ is the mass flow of the acid solution.

In the reactor model, window (quartz), steel shields (X6CrNiMoTi17-12-2) and insulation (alumina silicate, Alistra 1400) are additionally modeled. The model is split into two submodels, each accounting for the solid and the fluid phase, respectively. The boundary conditions are depicted in figure 5.19. Dual cell boundary describes the temperature and heat flux exchanged between the fluid and the solid phases. The convective heat boundary at the window is described by Nu correlations for vertical flat plates at constant temperature, while the convective heat transfer at the insulation is described by Nu correlations for horizontal cylinders [33, 34]. The wavelength-averaged reflectivity of the gold-covered aperture is assumed to be 0.98 [77].

Mesh and iteration convergence studies are performed to assure the solutions' independence of the mesh and the iteration progress. To reduce computational

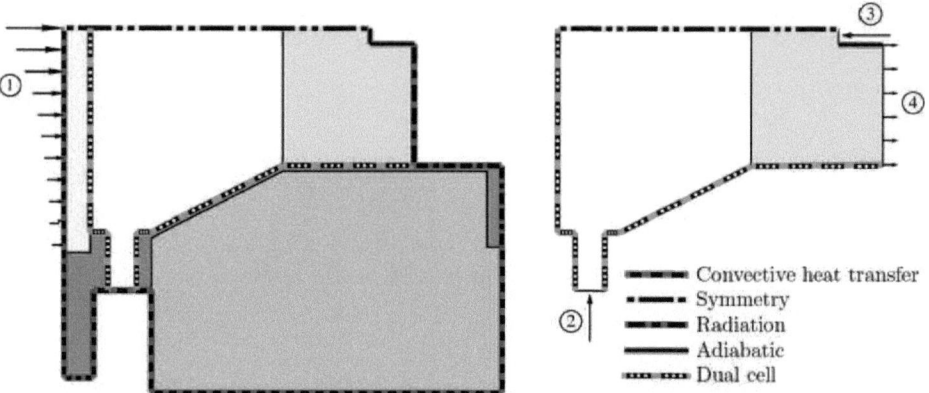

Figure 5.19: Boundary conditions for the external walls, used in the reactor model composed of solid domain (left, including vacuum domain between the window and the foam) and fluid domain (right). (1) solar irradiation inlet, (2) carrier gas inlet, (3) sulfuric acid inlet, and (4) outlet.

expense, the domain in the reactor model is divided in twelve symmetrical elements. This can only be done when the three acid inlets are combined into one central acid inlet. The final mesh of the reactor model consists of $80 \cdot 10^3$ and, for the foam model, of $180 \cdot 10^3$ tetrahedral elements. $40 \cdot 10^3$ iterations are used for convergence. The minimal discretization for the DO model is determined to be 3 in the two angular directions.

5.3.2 Comparison to experiment

Temperature measurements from the experimental campaigns performed at the solar furnace in Cologne are used for comparison with the model. A foam of 90 mm diameter and 40 mm thickness was used. Figure 5.20.a shows a cutting plane through the middle of the foam ($z = 0.02$ m), indicating the position of the thermocouples used for the temperature measurements, T_1–T_6. Temperature at position 7, T_7, is measured at the outlet of the reactor. Measured temperatures, power input and acid solution inflow are shown in figure 5.20.b. The temperature and power input measurements are averaged over each run (beginning with the start of the acid solution inflow and ending with the stop of the acid solution inflow), resulting in a steady state temperature and an averaged power input. Figure 5.21 compares the measured and the simulated temperature results for the experimental run, with total acid solution inflow of 2 ml/min (50 wt%

5.3. Continuum model

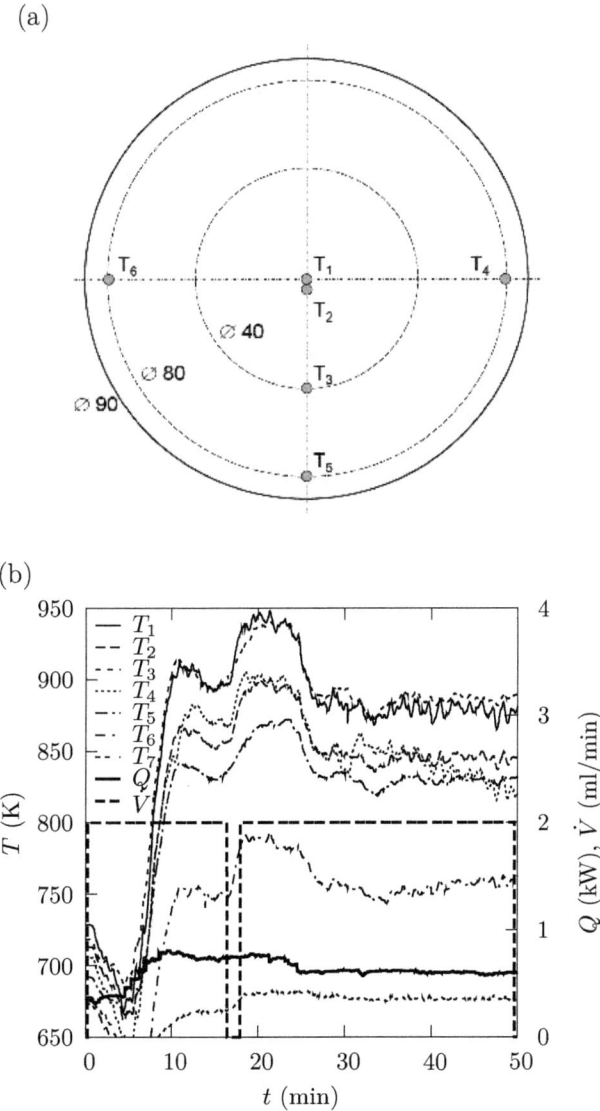

Figure 5.20: (a) Cutting plane through the middle of the foam, indicating the location of the thermocouples used for temperature measurements, and (b) recorded temperatures (left axis), acid solution inflow, and power input (right axis) of the experimental run at DLR.

sulfuric acid and 50 wt% water), 0.259 cm³/min carrier gas inflow, and 0.629 kW solar input power. The variations in the experimental data correspond to the standard deviations of the measured temperatures during the duration of the experiment.

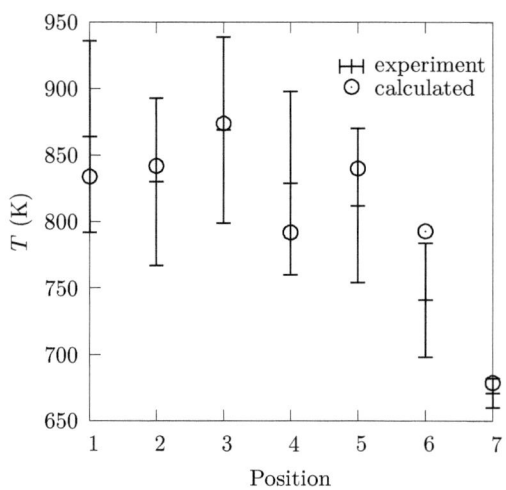

Figure 5.21: Comparison of the experimentally obtained and the numerically determined temperatures at the positions indicated in figure 5.20.

The discrepancies observed between the experimentally measured and numerically calculated temperatures at positions 4 and 6 are associated with an asymmetric angular distribution of the incident radiation. The difference observed at position 5 is associated with degradation in insulating material and, consequently, higher heat losses in the experiment.

5.3.3 Reference case

The previously described case will serve as the reference case for the following parameter study. The temperature distributions in the yz-plane (through the middle inlet tube, at $x = 0$) for the two phases are shown in figure 5.22. The carrier gas flowing in from the front ($z = 0.04$ m) at 300 K reaches the solid temperature within less than 5 mm. A small fraction's temperature of the fluid phase at the acid solution inlet drops below the evaporation temperature of the acid, indicating that the acid solution needs a finite amount of time to reach a hot surface and to be evaporated and decomposed. The temperature distributions at the reactor's foam outlet are shown in figure 5.23 for the solid and the fluid

5.3. Continuum model

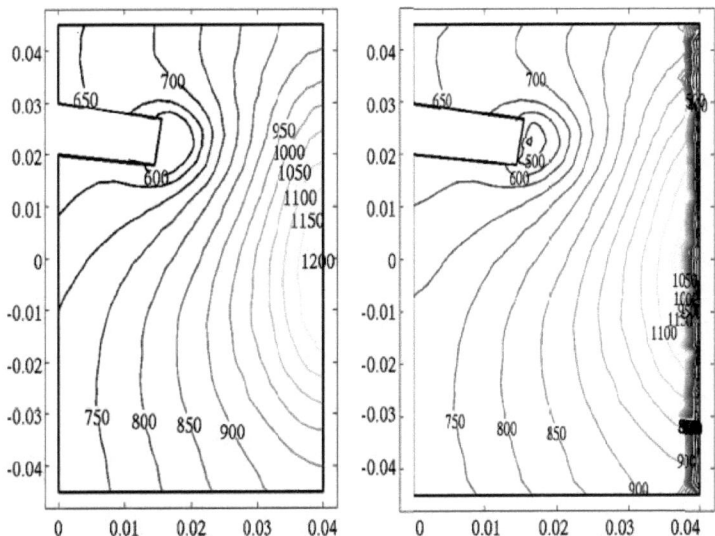

Figure 5.22: Temperature distribution in the yz-plane (through the middle inlet tube) for the solid phase (left) and the fluid phase (right).

phases. The lower temperatures at the top of the foam are explained with the position of the acid solution inlets. The solid temperatures are slightly lower because of the energy needed for the evaporation and decomposition reactions, solely delivered by the solid matrix.

The solid-to-fluid volumetric heat transfer is shown in figure 5.24. Three different lines, parallel to the z-axis (at $x = 0$), are shown: in the upper part of the foam ($y = 0.023$ m), in the middle of the foam ($y = 0$ m) and in the lower part of the foam ($y = -0.023$ m). In the upper part of the foam the line starts at $z = 0.015$ m, which corresponds to the position of the acid solution inlet of the middle tube. At the carrier gas inlet ($z = 0.04$ m), heat is transferred from the solid matrix, heated by the solar irradiation, to the fluid. At the acid solution inlet the solid matrix is rapidly cooled because of the instantaneous evaporation and decomposition of the acid solution. Heat is transferred from the hot carrier gas and the hot acid solution to the solid matrix. At the outlet, the solid matrix and the fluid are nearly at the same temperature, still a small portion of heat is transferred from the heated fluid phase to the solid phase.

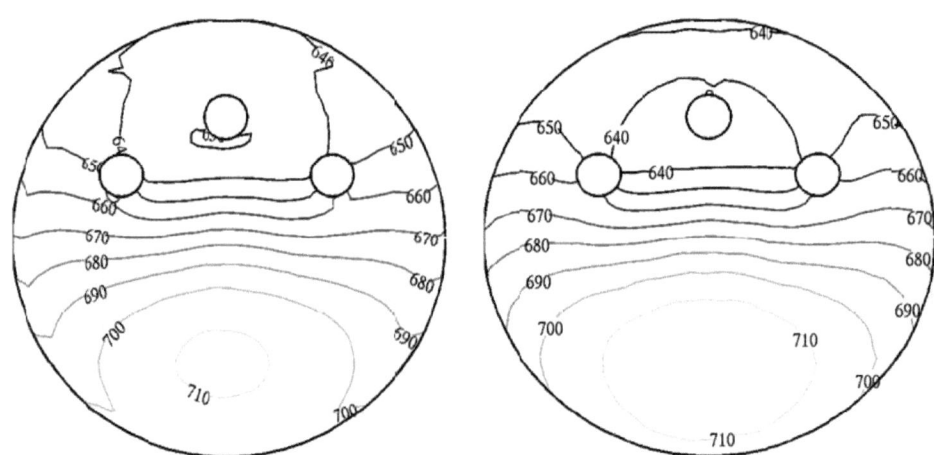

Figure 5.23: Temperature distribution at the outlet for the solid phase (left) and for the fluid phase (right).

5.3.4 Optimization

For comparison, two reactor efficiencies are defined. The energetic efficiency, which is defined as sensible heat, including evaporation, and reaction enthalpy for the decomposition reaction:

$$\eta_{en} = \frac{\dot{m}_{acid}\left(X_{evap}h_{acid,g,T_{out}} + (1-X_{evap})h_{acid,l,T_{out}} - h_{acid,l,T_{in}}\right)}{A_{in}q''_{in}} + \frac{\dot{m}_{N_2}(h_{N_2,T_{out}} - h_{N_2,T_{in}})}{A_{in}q''_{in}}, \quad (5.20)$$

where

$$h_{acid,g,T_{out}} = x_{H_2SO_4}\sum_i \frac{M_i}{M_{H_2SO_4}} h_{i,g,T_{out}} + (1-x_{H_2SO_4})h_{H_2O,g,T_{out}}, i = SO_3, H_2O, \quad (5.21)$$

and

$$h_{acid,l,T_j} = \sum_i x_i h_{i,l,T_j}, j = \text{in, out}; i = H_2SO_4, H_2O. \quad (5.22)$$

The chemical efficiency, defining the energy used solely for evaporation and decomposition:

$$\eta_{chem} = \frac{\dot{m}_{acid}X_{evap}(\Delta h_{evap} + \Delta h_{decomp})}{A_{in}q''_{in}}. \quad (5.23)$$

5.3. Continuum model

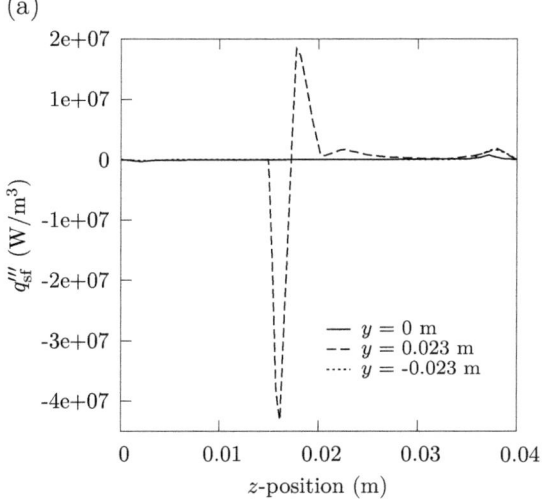

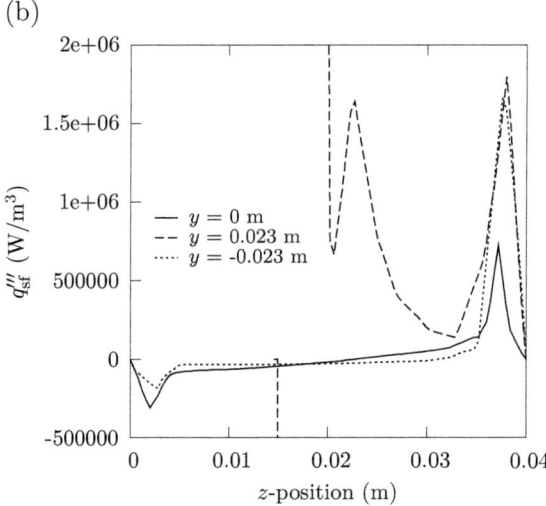

Figure 5.24: Solid-to-fluid volumetric heat transfer along different lines, parallel to the z-axis, in the foam (a). A zoom is depicted in (b). The acid solution inlet of the middle tube is located at $y = 0.023$ m and $z = 0.015$ m.

The chemical conversion is defined as

$$X_{\text{evap}} = \frac{\dot{m}_{\text{acid,g}}}{\dot{m}_{\text{acid,in}}}. \tag{5.24}$$

Area-averaged temperatures on the front side and the back side of the porous absorber, and volume-averaged temperature within the absorber are also given for comparison.

Parameter studies focusing on three areas are conducted for process optimization: (i) operational conditions, (ii) reactor design, and (iii) foam morphology.

Operational conditions

Total acid solution inflow is varied between 1 and 12 ml/min. Efficiencies are depicted in figure 5.25.a and temperatures in figure 5.25.b. Increasing the acid solution inflow leads to a decrease of the absorber temperature because more energy is needed for the heating, evaporation and decomposition. For an acid solution inflow greater than 5 ml/min, the minimum temperature on the outlet drops below the evaporation temperature of the acid solution and consequently the chemical conversion is incomplete. Nevertheless, the energetic and the chemical efficiency increases up to an acid solution inflow of 6 ml/min with peak values of 57% and 32%, respectively.

The solar power input is varied between 75 and 1500 W. In figure 5.26 the influence of the solar power input on the reactor efficiencies, the chemical conversion and the absorber temperatures are depicted. Although a radiation power below 300 W leads to incomplete conversion of the acid solution, the energetic and chemical efficiencies increase. The reactor performance peaks at a solar power input of 150 W with a maximum energetic and chemical efficiencies of 73% and 45%, respectively. Reducing the input power further prevents evaporation, and the chemical conversion drops to zero. As seen in figure 5.26, the reduction of the power input decreases the absorber temperature.

For both operational conditions the energetic efficiency peaks at the transition region from complete conversion to partial conversion. At this operating point, most of the diluted sulfuric acid is evaporated and decomposed. The excess power is minimal. This leads to a absorber temperature as low as possible without diminishing the chemical conversion to a too great extent. Lower temperatures reduce the re-radiation losses and therefore increase the reactor performance.

5.3. Continuum model

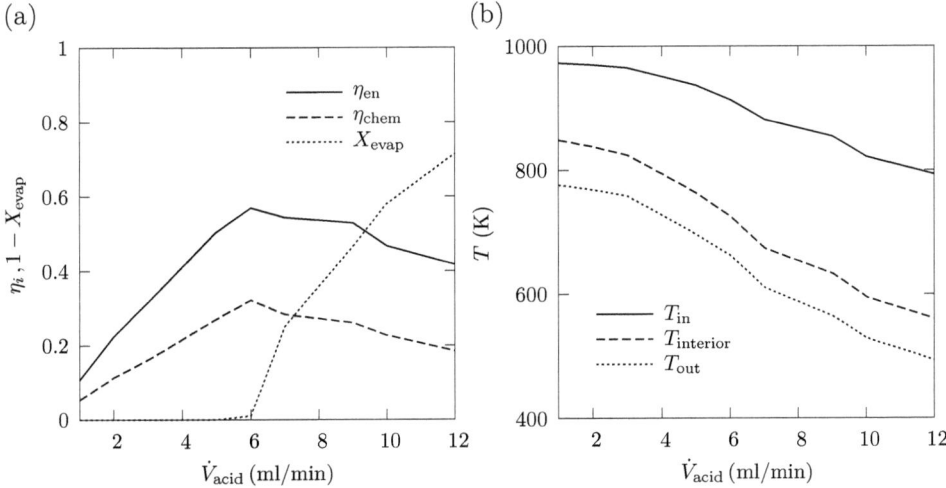

Figure 5.25: Energy and chemical efficiency, and evaporation conversion as a function of total acid solution inflow (a), and inlet, interior and outlet temperature as a function of total acid solution inflow (b).

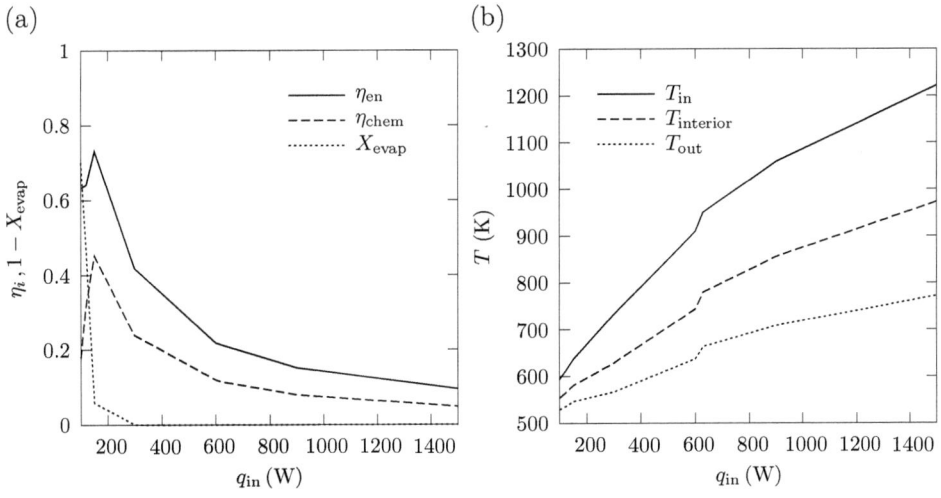

Figure 5.26: Energy and chemical efficiency, and evaporation conversion as a function of solar input power (a), and inlet, interior and outlet temperature as function of solar input power (b).

Reactor design

The length of the porous absorber and the diameter of the central acid solution inlet are varied in order to investigate on the influence of reactor design decisions.

The length is varied between 4 and 6.5 cm. Minor changes in efficiencies and conversion are observed when increasing the absorber length. The energetic and chemical efficiencies stay constant at 21% and 11%, respectively, and full evaporation conversion is achieved. Increasing the absorber length leads to an increase in the front side temperature of the absorber by 50 K due to the limitations in the heat conduction through the foam.

The diameter of the tube through which the acid solution flows into the reactor is varied between 14 and 24 mm. Its influence on efficiencies, conversion and temperature is minimal. The energetic and chemical efficiencies stay constant at 21% and 11%, respectively, and full evaporation conversion is achieved. For both small and large diameters, the average absorber temperature increases by 20 K. Increasing the inlet diameter reduces the inlet velocity and, therefore, decreases the Nu number. On the other hand, decreased inlet diameter reduces the area of instantaneous evaporation, leading to larger local heat sinks.

In conclusion, the influence of the reactor design on performance is small compared to the influence of the operational conditions.

Foam design

Nominal pore diameter and porosity are varied in order to investigate on the influence of the foam morphology on reactor performance. Porosity is varied between 0.5 and 0.95, and the nominal pore diameter between 1 and 10 mm. Again, the influence is minimal compared to the influence of the operational conditions.

Changing the nominal pore diameter while assuming that there are no significant changes in pore size distribution (morphology) and porosity, has several effects on the transport properties of the foam. Increasing the pore size decreases the absorption of the foam, allowing the radiation to penetrate deeper into the foam, heating the foam more uniformly and reducing the re-radiation losses. Decreasing the pore size decreases permeability, and consequently increases the pressure drop in the foam, and increases Dupuit-Forchheimer coefficient. This leads to an increase in residence time and tortuosity which allows heating up the fluid to higher temperatures. Increasing pore size leads to higher Nu numbers, allowing for a more efficient convective heat transfer. Nevertheless, these two processes are not dominant. Changes in pore size (at constant porosity and

5.4. Conclusions

unchanged morphology) do not affect the effective conductivity of the foam. The absorber pore size has only a minor influence on the reactor efficiency and the temperature distribution in the absorber foam. The energetic and chemical efficiencies stay constant at 21% and 11%, respectively, and full evaporation conversion is achieved. As seen in figure 5.27.a, increasing the pore size from 1 mm to 10 mm leads to a temperature decrease of only 1%.

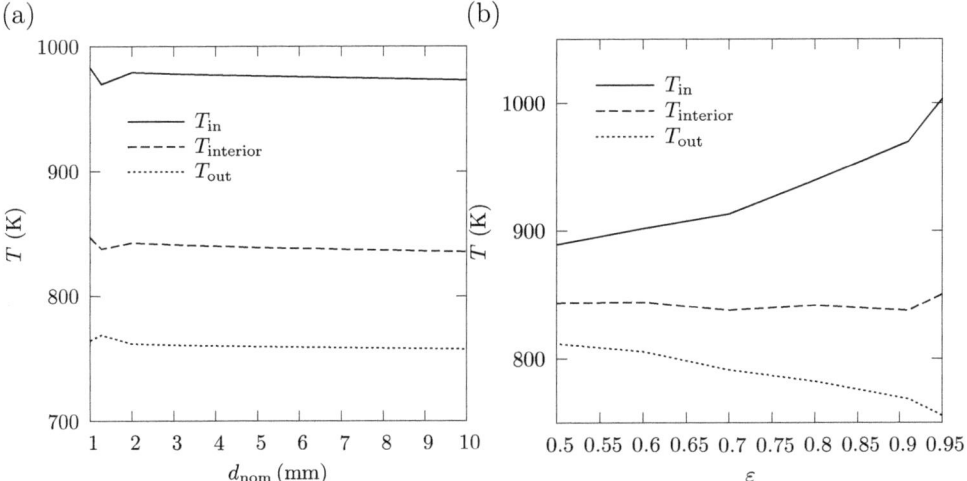

Figure 5.27: Inlet, interior and outlet temperature as a function of nominal diameter of the foam (b), and as a function of foam's porosity (b).

In figure 5.27.b, the temperature of the absorber for different porosities is shown. A decrease of the porosity leads to a more uniform temperature distribution in the absorber. This is due to the fact that reduction of the porosity improves the thermal conductivity of the absorber and hence a more equally heating of the porous absorber is achieved.

5.4 Conclusions

The effective heat and mass transfer properties of a 20 ppi non-hollow RPC foam made of SiSiC, whose exact 3D geometry was determined by CT, has been numerically computed. Computed porosity was 0.91 and compared well to the experimentally measured value of 0.90 ± 0.02. Computed specific surface was 1367 m^{-1} and increased by 20% when increasing the scan resolution by factor 2 as smaller surface irregularities were resolved. Computed pore size distribution showed a sharp peak of approximately $0.5d_{\text{nom}}$ and a mean diameter of $1.3d_{\text{nom}}$. REV determined by porosity, extinction coefficient, and conductivity calculations on subsequently growing volumes was 87.8 mm^3. Radiative properties were determined from the extinction length and cosine of incidence distributions by applying the collision-based MC method. The computed extinction coefficient of 431 m^{-1} agreed quantitatively to the experimentally measured estimated by measuring the transmitted radiative flux with a spectroscopy system. Computed scattering functions showed a large backward scattering peak for diffusely reflecting surfaces and isotropic scattering behavior for specularly reflecting surfaces. The scattering coefficient was a function of the surface reflectivity and determined to be 63 m^{-1}. The effective conductivity was calculated by solving the heat conduction equation within both phases by FV and fitted to a combination of parallel and serial slab models. For $k_{\text{f}}/k_{\text{s}} < 10^{-4}$, k_{e} remained constant and approximately $0.022k_{\text{s}}$. The heat transfer coefficient was calculated by solving the continuity, momentum, and energy governing equations within the fluid phase by FV. A Re and Pr dependent Nu correlation of the form $\text{Nu} = 6.820 + 0.198\text{Re}^{0.788}\text{Pr}^{0.606}$ was fitted (RMS = 0.817). This correlation was strongly dependent on morphology. Computed permeability and Dupuit-Forchheimer coefficient, determined based on the pressure and velocity distribution within the fluid, were $K = 5.67 \cdot 10^{-8}$ m^2 and $F_{\text{DF}} = 519.0$ m^{-1}, compared well to the values found by applying different models available in the literature. Tortuosity distribution calculations resulted in a mean tortuosity of 1.07. Obviously, the mean residence time decreased with increasing Re. Neglecting molecular dispersion, a Re dependent function of the dispersion tensor was obtained by comparing the calculated spatial displacement distribution of streamlines within the foam to a Gaussian distribution.

The CT-based methodology was able to accurately account for the morphology of complex porous media, and, when coupled to Monte Carlo and CFD numerical techniques, provided pore-level solutions of the energy and fluid flow governing equations. The limits of applicability of non-combined conduction-convection-radiation computations remain to be determined.

The effective transport properties were then used in the continuum-scale heat and mass transfer model, which, in turn, was used for the design and optimization of the solar evaporation/decomposition reactor. The continuum model was compared with experimental temperature measurements conducted obtained at the solar furnace of DLR at Cologne. Good agreement was observed. In a subsequent parameter study, operational conditions, reactor design and foam morphology were varied and their influence on reactor efficiencies, evaporation

5.4. Conclusions

conversion, and temperatures was investigated.

What influenced the reactor performance the most was varying the total acid solution mass flow rate and the solar power. In order to achieve an optimum energetic efficiency, the interaction and coordination between these two operational conditions were essential. The energetic and chemical efficiencies peaked for both operational conditions at evaporation conversion between 1 and 0.9. This was the transition region from complete conversion to partial conversion, which indicates that a complete conversion was not necessarily the aim to improve the reactor performance with regard to the efficiencies. With a solar power input of 650 W, a diluted sulfuric acid inflow of 6 ml/min was required to reach the optimum of energetic and chemical efficiencies (57% and 32%, respectively). For an acid solution inflow of 2 ml/min, the reactor efficiency peaked at a solar power input of 150 W with maximum energetic and chemical efficiencies of 73% and 45%, respectively. The influence of the absorber porosity and the pore size diameter on the reactor efficiencies was minor. Reduced porosity led to a more uniform temperature distribution within the absorber.

In a further approach the continuum model developed for the solar evaporator/decomposer reactor can be extended to a three-phase model to accurately account for the evaporation process. This allows for an in-depth understanding of the evaporation process in the porous media and for further optimization of the reactor.

Chapter 6
Anisotropic ceramic foams[1]

In this chapter an anisotropic ceramic foam made of ceria is analyzed for its effective heat and mass transfer properties. It is considered as radiation absorber, reactant and reaction site for a two-step thermochemical cycle for splitting of H_2O and/or CO_2 via redox reactions. Highly concentrated solar energy is used to drive the reaction.

High-resolution X-ray tomography is employed to obtain its exact 3D geometrical configuration, which in turn is used in discrete-scale numerical calculations for determining the morphological and directional effective heat/mass transport properties, namely: porosity, specific surface area, pore size distribution, extinction coefficient, thermal conductivity, convective heat transfer coefficient, permeability, Dupuit-Forchheimer coefficient, and tortuosity and residence time distributions.

Few tailored foam designs for enhanced transport properties were examined by means of adjusting morphologies with artificial ceria samples composed of bimodal distributed overlapping transparent spheres in an opaque medium.

CT-based approaches have been used for the determination of the fibre orientation [220] and anisotropic permeability [45] in fibrous materials, the extinction coefficient and scattering phase function in an anisotropic ceramic foam [238], permeability [169], diffusivity [147] and tortuosity [146] in anisotropic rock samples.

---

[1] Material from this chapter has been published in: S. Haussener, and A. Steinfeld. Effective heat and mass transport properties of anisotropic porous ceria for solar thermochemical fuel generation. *Materials*, 5:192–209, 2011. [84]

6.1 Metal oxide/metal cycles with the non-stoichiometric CeO_2/Ce redox pair

In the non-stochometric CeO_2/Ce-based (generally described by $CeO_{2-\delta}/Ce_{2-\delta-x}$) water-splitting cycles for the production of hydrogen, $CeO_{2-\delta}$ is partially reduced in the endothermic high-temperature step at temperatures around 1800 K

$$CeO_{2-\delta} = CeO_{2-\delta-x} + \frac{x}{2}O_2. \tag{6.1}$$

In the exothermic oxidation step the $CeO_{2-\delta-x}$ is partially oxidized with steam (or CO_2) to produce H_2 and $CeO_{2-\delta}$ at approximately 1100 K,

$$CeO_{2-\delta-x} + xH_2O = CeO_{2-\delta} + xH_2. \tag{6.2}$$

$CeO_{2-\delta}$ is re-used in the high-temperature reduction reaction. The net reaction is splitting of water. $CeO_{2-\delta}/Ce_{2-\delta-x}$-based thermochemical cycles allow the production of H_2 over an oxygen non-stoichiometric change within the framework of a fixed crystal structure eliminating the losses due to partial product gas recombination and consequently the need for energy and resource intense gas quenching. Since melting ($T_{\text{melt},CeO_{2-\delta}} \approx 2800$ K) and sintering do not occur during reactions (6.1) and (6.2), full accessibility of the reaction sites is guaranteed. Additionally, it shows rapid fuel production kinetics and high selectivity [31]. Alternatively, this process allows for the production of syngas if a combination of H_2O and CO_2 is used as the oxidant in reaction (6.2). In the presence of a base-metal catalyst, simultaneous production of methane can be achieved [32].

Water splitting by the use of ceria-based materials has been investigated recently. Solid solutions of metal oxides (MnO, NiO, CuO, Fe_2O_3, ZrO_2) and CeO_2 are used for solar driven thermochemical water splitting with enhanced redox reactions reported compared to pure ceria-based cycles [100, 140, 2]. Feasibility of solar ceria-based system used for water splitting is shown in [1, 50].

Solar reactor design proposed for the $CeO_{2-\delta}/Ce_{2-\delta-x}$-based water splitting reaction is a cavity-type receiver as shown in figure 6.1, allowing for the reduction and oxidation step in the same reactor. Solar radiation enters through a quartz window and is further concentrated by a compound parabolic concentrator (CPC). A ring of porous ceria bricks or felt absorbs the radiation, enhances the distribution of the gasifying agent, and serves as reactant and reaction site [50]. To ensure efficient heat and mass transport to and from the surface reaction sites and thereby maximize the kinetics of these reactions, it is necessary to utilize the ceria in porous form. The inert carrier gas or the H_2O

6.2. Real samples

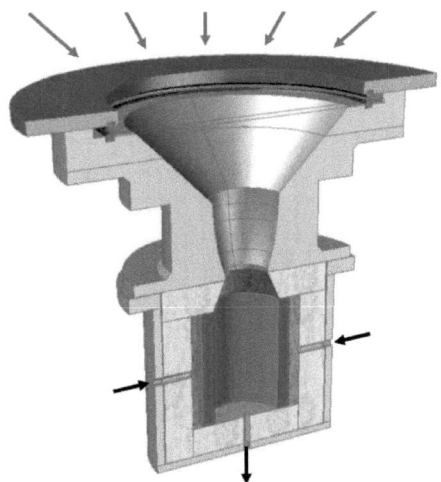

Figure 6.1: Schematic of the porous ceria-based water-splitting reactor [50].

and/or CO_2 flows enter the reactor at four inlets from the side. The product gas leave the reactor at the bottom.

6.2 Real samples

6.2.1 Computed tomography

Three types of porous ceria, each of approximately 65% porosity and prepared using graphite as a sacrificial pore-former, were used for the analysis. Graphite (Alfa Aesar 40769 and 10129, see figure 6.2.a and b) and ceria (Alfa Aesar 11328) powders are mixed in ethanol for 10 min, dried, uniaxially pressed in a 13 mm-diameter die at 220 MPa for 2 min, and sintered at 1500 °C for 5 hours (5 and 1 K/min heating and cooling rate, respectively). The three samples differ by their volumetric fractions of the two graphite powder types (Alfa Aesar 10129 and 40769). Laser scattering (HORIBA LA-950) is used for the determination of the particle-size distribution of the sacrificial pore-former, plotted in figure 6.2.c. Note that these distributions are qualitative as particles are not spherical, which is a base assumption in Mie scattering theory [16] applied for the recalculation of the size distributions. Experimentally determined mean diameters are 28 μm and 270 μm for the Alfa Aesar 10129 and 40769, respectively. The inherent morphological anisotropy of the graphite powders and the directional processing creates porous ceria with structural anisotropy. Effective properties of the anisotropic sample are calculated along the three principal directions of the sample: the direction of uniaxial pressing (z-direction) and the two orthogonal directions (x- and y-direction).

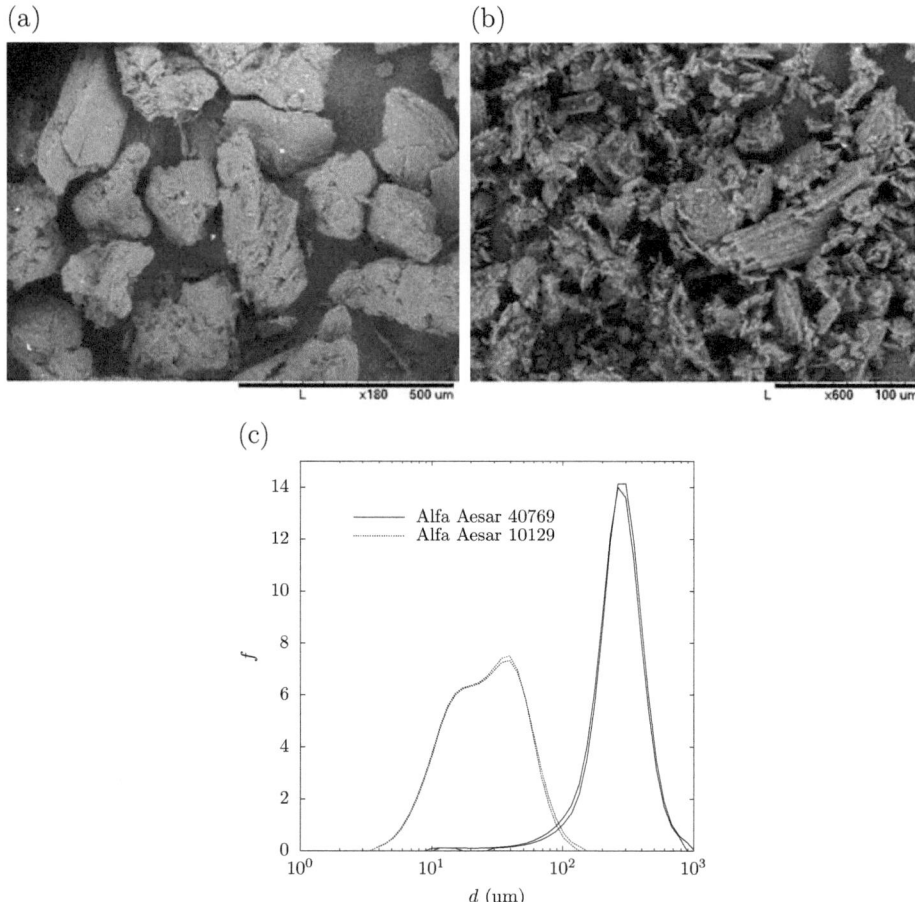

Figure 6.2: SEM pictures (HORIBA TM1000) of the graphite powder, Alfa Aeasar 40769 (a) and Alfa Aesar 10129 (b), and volume-based particle-size distributions, f, of two random samples of each graphite powder type (c).

HRCT is obtained through synchrotron radiation of the TOMCAT beamline of SLS at PSI [208, 200] with the following operating conditions: 36 keV photon

6.2. Real samples

energy, 400 µA beam current, 100 µm Al and 10 µm Fe and 40 µm Cu filter, 2.1 s exposure time, and 1500 projections. The data's voxel size is 0.37 µm, and FOV investigated is 0.76 x 0.76 x 0.76 mm^3. A CT scan and a 3D rendered picture of sample no. 2 are shown in figure 6.3. The discrete absorption values obtained by HRCT are linearly interpolated in 3D to obtain a continuous representation of the phase boundary. The tomographic data is segmented via the mode method.

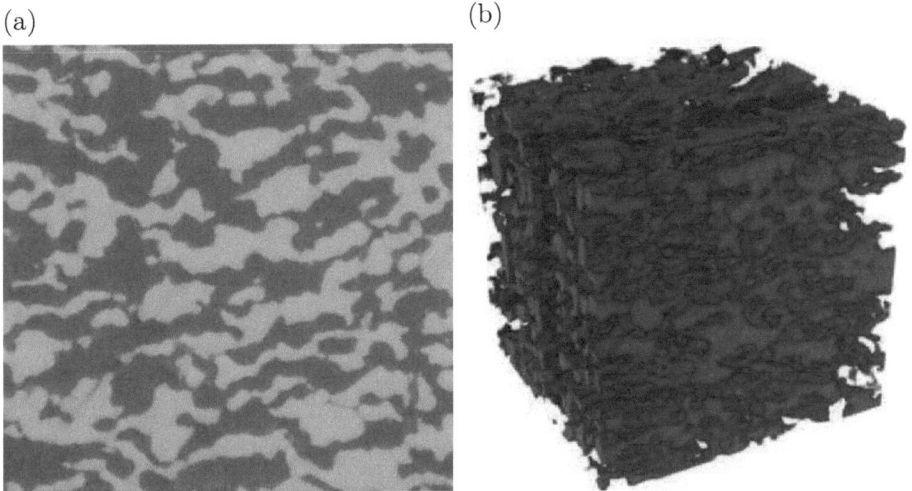

Figure 6.3: CT scan (a) and 3D rendering (b) of the porous ceria foam with edge length of 376 µm.

6.2.2 Morphological characterization

Calculated sample porosities and specific surface areas, based on two-point correlation functions, are given in table 6.1. The underestimation of the calculated porosities compared to the experimentally determined porosities, obtained by weight measurements, is due to the limited resolution of the tomography data.

The pore-size distribution based on openings with spherical structuring elements is shown in figure 6.4.a. As expected, sample no. 1 shows the biggest fraction of the largest pores. Calculated mean, mode, and median diameter of the obtained pore-size distributions are given in table 6.1.

l_{REV} is determined based on porosity calculations on subsequently growing volumes. The calculated porosity as a function of l_{REV} for 20 randomly chosen locations in the sample is shown in figure 6.4.b. l_{REV} for the three samples inves-

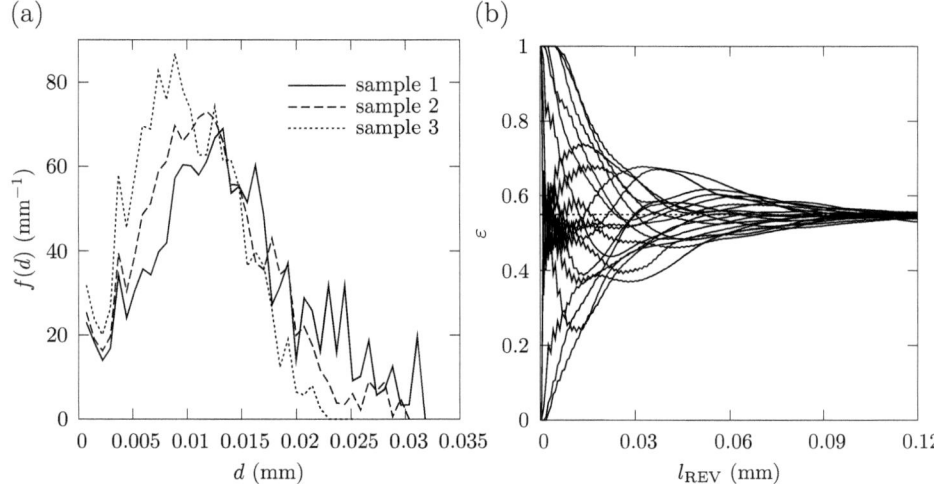

Figure 6.4: Opening pore-size distribution of the three ceria samples (a) and porosity calculated for sample no. 3 in 20 subsequently growing volumes (b). The dotted horizontal line in (b) indicates $\varepsilon_{\text{num}}^2$ to which the values converge.

tigated are 0.11 mm, 0.10 mm, and 0.06 mm, respectively, assuming a porosity band of ±0.06 to be sufficient. For the following calculations, the minimum sample size is given by REV.

The mean intercept length for sample no. 1 is shown in figure 6.5.a. Mean intercept length is defined as the average distance between two solid-fluid boundaries [197]. θ and φ represent the polar and azimuthal angles in spherical coordinates. Scattering in the data is due to statistical variation in the morphology. The intercept length along the x- and y-axis ($\theta = 90°$ and $\varphi = 0°$, and $\theta = 90°$ and $\varphi = 90°$, respectively) is 25 to 35% larger because, due to uniaxial pressing, the pores in x- and y-directions are elongated and channel-like structures develop, while the pores in z-direction are squeezed. This is exemplary shown in figure 6.5.b by the three orthogonal planes of sample no. 1 and the two black bars indicating elongated pores and channels evolving along the x- and y-axis.

6.2.3 Heat transfer characterization

Radiation

The solid volume of the ceria foam is assumed to be opaque, while the void volume is assumed to be transparent ($\sigma_{\text{s,d}} = \kappa_{\text{d}} = 0$), as the gas absorption coefficients of the reacting gases are orders of magnitude smaller than the ab-

6.2. Real samples

Table 6.1: Numerically and experimentally determined porosity, numerically determined specific surface, mean, mode, median and hydraulic diameter of the three porous ceria samples.

Sample no.	1	2	3
ε_{num}	0.51	0.56	0.55
ε_{ex}	0.65 ± 0.01	0.65 ± 0.01	0.65 ± 0.01
A_0 (mm^{-1})	672	706	675
d_{m} (µm)	13.7	11.9	20
d_{mode} (µm)	13.3	11.8	8.9
d_{median} (µm)	13.1	11.5	9.7
d_{h} (µm)	3.0	3.2	3.3

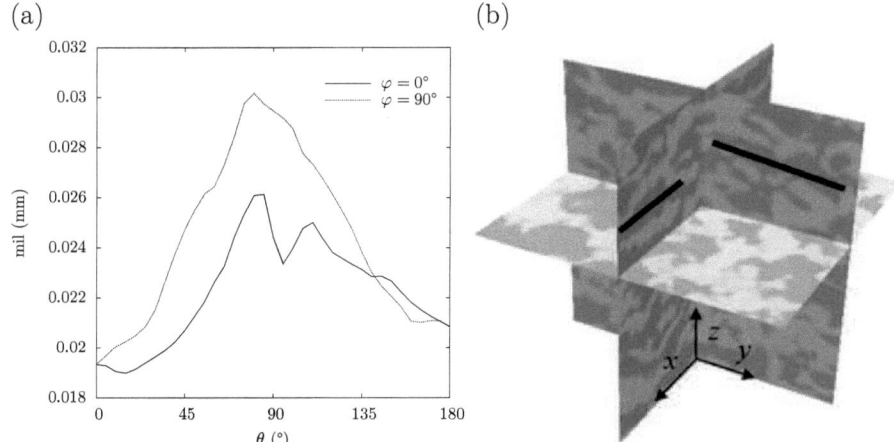

Figure 6.5: Mean intercept length for ceria sample no. 1 as a function of θ at $\varphi = 0$ and 90° (a), and three orthogonal planes of sample no. 1 with black bars indicating the direction of the elongated pores (b).

sorption of radiation at the solid-fluid boundary. For example, the Planck mean absorption coefficient of CO at 1000 K and 1 atm is 2.1 m^{-1}, calculated based on the HITRAN2004 database [183]. The governing equations and the methodology to determine the radiative properties of a two-phase media composed of an opaque and a transparent phase are given in chapters 2 and 3. A sample of 0.37 x 0.37 x 0.37 mm^3, corresponding to 1000 x 1000 x 1000 voxels, is investigated and $N_{\text{ray}} = 6 \cdot 10^6$.

Table 6.2: Numerically determined extinction coefficients of the three porous ceria samples in the three principal directions.

Sample no.	1	2	3
β (m^{-1}), x-direction	30003 ± 8282	38143 ± 6277	45173 ± 4665
β (m^{-1}), y-direction	31757 ± 7067	35042 ± 4546	46277 ± 6485
β (m^{-1}), z-direction	69018 ± 14735	65665 ± 9809	74835 ± 15022

The extinction coefficients along the x-, y-, and z-directions are listed in table 6.2. These values are consistent with the correlation by Hendricks [87]. The calculated extinction coefficients show the same order of magnitude as those estimated by transmittance measurements of similar ceria foams ($\varepsilon = 0.72$, graphite and starch pore-formers) at low wavelengths where the opacity assumption of solid ceria holds [116]. As expected for small pore dimensions, the porous ceria foam behaves as a nearly opaque medium. Due to uniaxial pressing, the pores in x- and y-directions are elongated, while the pores in z-direction are squeezed and, consequently, lead to shorter extinction path lengths and larger extinction coefficients. In addition, the increasing fraction of smaller pores from sample no. 1 to no. 3 leads to increased extinction. This trend is less pronounced in the z-direction because circumferentially squeezing of an oblate by factor 2 leads to its elongation by factor 4. Note that the uniaxial pressing is not identical for the three samples, resulting in different grades of anisotropy.

Conduction

The governing equations and methodology to determine the effective heat conductivity in a connected porous foam are described in chapters 2 and 3. A sample of 0.37 x 0.37 x 0.37 mm^3, corresponding to 1000 x 1000 x 1000 voxels, is investigated. Grid convergence is achieved with a mesh element size of 21.5 µm. Calculated effective thermal conductivity normalized by the solid conductivity as a function of fluid to solid conductivity, k_f/k_s, is shown in figure 6.6. Also included in the graph are the maximum and minimum conductivities possible in a regular ordered two-phase media, described by serial and parallel slab models [102], with $\varepsilon = 0.51$. As can be expected, the conductivity of the foam lies within these boundaries.

As the fluid conductivity decreases, the influence of the sample morphology becomes more pronounced while conduction in the fluid is less important. Due to the uniaxial pressing, the structures in the x- and y-directions are aligned

6.2. Real samples

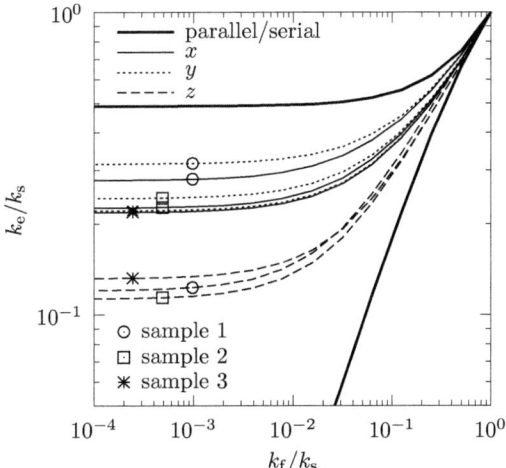

Figure 6.6: Effective normalized conductivity of the three samples as functions of k_f/k_s in the three directions and for parallel and serial slab models at $\varepsilon = 0.51$.

Table 6.3: Fitting parameter a_1 of eq. 6.3.

Sample no.	1	2	3
a_1, x-direction	0.427	0.443	0.506
a_1, y-direction	0.354	0.482	0.512
a_1, z-direction	0.751	0.740	0.705

more parallel to the heat flux and exhibit enhanced conductivities compared to that in the z-direction. k_f/k_s lies nearer to the conductivities predicted by the parallel slab model. Again, samples no. 2 and no. 3 show a less pronounced anisotropy compared to sample no. 1. An analytical description of the effective thermal conductivity as a function of k_f/k_s is derived by fitting the numerically determined results to a model described by a summation of serial and parallel slab models,

$$\frac{k_e}{k_s} = a_1 \frac{\frac{k_f}{k_s}}{\varepsilon \left(1 - \frac{k_f}{k_s}\right) + \frac{k_f}{k_s}} + (1 - a_1) \left(\varepsilon \frac{k_f}{k_s} + 1 - \varepsilon\right). \tag{6.3}$$

The fitted parameter a_1 is given in table 6.3 for three samples and in the three directions. The RMS of the fitting was less than 0.165.

Convection

The governing equations and methodology to determine the heat transfer coefficient are described in chapter 2 and 3. A sample of 0.33 x 0.33 x 0.17 mm^3, corresponding to 900 x 900 x 450 voxels, is investigated. Convergence of the numerical calculations is achieved for a terminal residual RMS of the iterative solution below 10^{-5} and for a maximal mesh element length of 3 μm. The resulting meshes have tetrahedral element numbers between 50 and 150·10^6.

Calculated Nu as a function of Re (based on $d = 10$ μm) in the range 0.1 and 100, and for Pr = 0.1 and 1, are shown in figure 6.7 in the three directions. They are fitted to a correlation of the form given by eq. (3.23), with the constants a_1 to a_4 given in table 6.4. The RMS of the fitting was less than 0.6. The heat transfer coefficient increases along the z-direction because of the more tortuous path for fluid flow, increasing the accessible surface area for fluid-solid heat exchange. This trend is most pronounced for sample no. 1. Values for Nu lay above those experimentally measured for ceramic foams [237], but within the range of those for packed beds [225], see eq. (7.20).

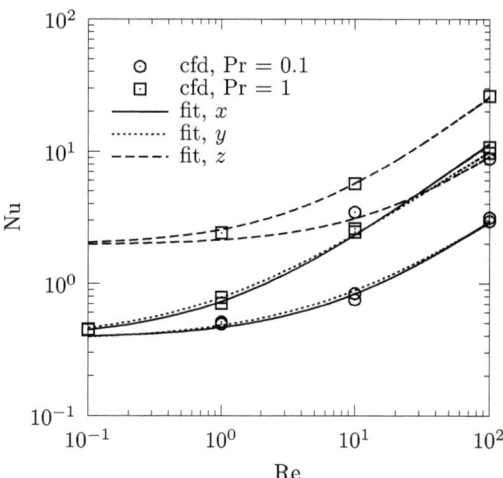

Figure 6.7: Nu number as a function of Re and Pr numbers (points) and fit (lines) for sample no. 1 in the three directions.

6.2. Real samples

Table 6.4: Nu correlations for the three porous ceria samples in the three principal directions.

Sample no.	1	2	3
Nu, x-dir.	$0.38+0.35 Re^{0.75} Pr^{0.64}$	$1.09+0.50 Re^{0.69} P^{0.56}$	$0.82+0.64 Re^{0.66} Pr^{0.51}$
Nu, y-dir.	$0.37+0.41 Re^{0.68} Pr^{0.58}$	$0.75+0.53 Re^{0.68} Pr^{0.59}$	$0.81+0.68 Re^{0.65} Pr^{0.47}$
Nu, z-dir.	$1.96+0.60 Re^{0.80} Pr^{0.52}$	$1.28+0.65 Re^{0.75} Pr^{0.57}$	$0.97+0.97 Re^{0.67} Pr^{0.51}$

6.2.4 Mass transfer characterization

Permeability and Dupuit-Forchheimer coefficient

CFD at the pore level in the laminar fluid phase is applied to determine K and F_{DF}, as described in chapters 2 and 3. Π_{pg} for sample no. 1 is shown in figure 6.8.a as a function of Re (based on $d = 10$ µm) for the three directions. The Dupuit-Forchheimer term comes into play at Re > 0.5. Calculated K and F_{DP} are given in table 6.5 in the three directions. Channels in the x- and y-directions that evolved during the uniaxial pressing lead to smaller pressure gradients and, consequently, higher permeabilities. The Dupuit-Forchheimer term decreases as well, since less inertia forces apply to fluid particles flowing through the foam in the x- and y-directions in a less disturbed manner. This trend is again most pronounced for sample no. 1. K versus F_{DP} is shown in figure 6.8.b for the three samples and in the three directions. Comparison of the calculated K with estimates by the capillary model ($K = 2.03 \cdot 10^{-12}$ m^2, see eq. (7.25)) [102], hydraulic radius model ($K = 1.25 \cdot 10^{-12}$ m^2, see eq. (7.26)) [102], and fibrous bed model ($K = 0.86 \cdot 10^{-14}$ m^2, see eq. (7.27)) [113], and of the calculated F_{DP} with estimates by extended Ergun equation ($F_{DP} = 22.9 10^4$ m^{-1}, see eq. (7.31)) [126], for $\varepsilon = 0.65$ and $d = 10$ µm, show reasonable agreement, given the differences in morphology.

Dispersion tensor

The calculated dispersion tensor for sample no. 1 is given by

$$D_{\perp,x/y} = 8.32 \cdot 10^{-5} Re^{0.558}, \tag{6.4}$$

$$D_{\|,x/y} = 1.19 \cdot 10^{-3} Re^{0.487}, \tag{6.5}$$

for the x- and y-direction, while for the z-direction it is given by

$$D_{\perp,z} = 9.56 \cdot 10^{-5} Re^{0.562}, \tag{6.6}$$

$$D_{\|,z} = 2.00 \cdot 10^{-3} Re^{0.514}. \tag{6.7}$$

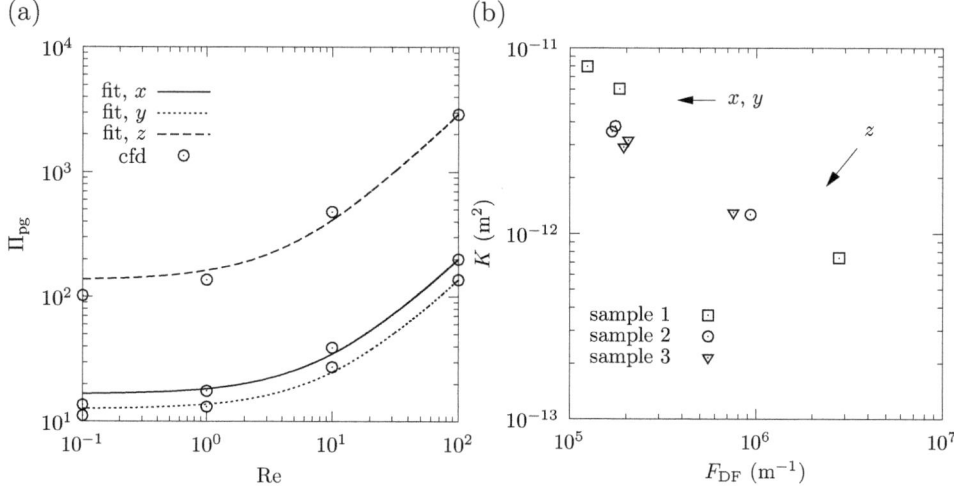

Figure 6.8: Normalized pressure drop as a function of Re number for the three directions in sample no. 1 (a), and calculated K versus F_{DP} for the three samples and in the three directions (b).

Tortuosity and residence time distributions

Tortuosity and residence time distributions, calculated for sample no. 1 and $l_{\mathrm{sample}} = 0.17$ mm, are shown in figure 6.9. Tortuosities in the x- and y-directions are smaller and show narrower peaks than the one in the z-direction, as the fluid is able to pass through the foam along the channels formed during the uniaxial pressing. Tortuosity along the z-direction decreases as Re increases because of the evolution of vortices in partially connected pores. The vortices block them and force the fluid to flow though more direct paths. Residence time distributions in the z-direction show a sharper peak than that in the two other directions, especially at large Re. At small Re, the distribution shows a tail due to fluid particles temporarily trapped in partially connected pores or death ends. Calculated mean tortuosities and residence times are given in table 6.6 for Re = 1.

6.3 Tailored foam design

Optimization of the thermochemical process and the associated solar reactor is closely related to enhanced heat and mass transport within the porous material. Our pore-level engineering approach in this regard is based on artificially gen-

6.3. Tailored foam design

Table 6.5: Calculated permeability and Dupuit-Forchheimer coefficient for the three samples in the three principal directions.

Sample no.	1	2	3
K (m^2), x-direction	$6.04 \cdot 10^{-12}$	$3.54 \cdot 10^{-12}$	$2.92 \cdot 10^{-12}$
K (m^2), y-direction	$7.97 \cdot 10^{-12}$	$3.80 \cdot 10^{-12}$	$3.03 \cdot 10^{-12}$
K (m^2), z-direction	$7.43 \cdot 10^{-13}$	$1.27 \cdot 10^{-12}$	$1.21 \cdot 10^{-12}$
F_{DP} (m^{-1}), x-direction	$18.4 \cdot 10^4$	$16.8 \cdot 10^4$	$19.4 \cdot 10^4$
F_{DP} (m^{-1}), y-direction	$12.4 \cdot 10^4$	$17.5 \cdot 10^4$	$19.3 \cdot 10^4$
F_{DP} (m^{-1}), z-direction	$278.9 \cdot 10^4$	$93.5 \cdot 10^4$	$74.9 \cdot 10^4$

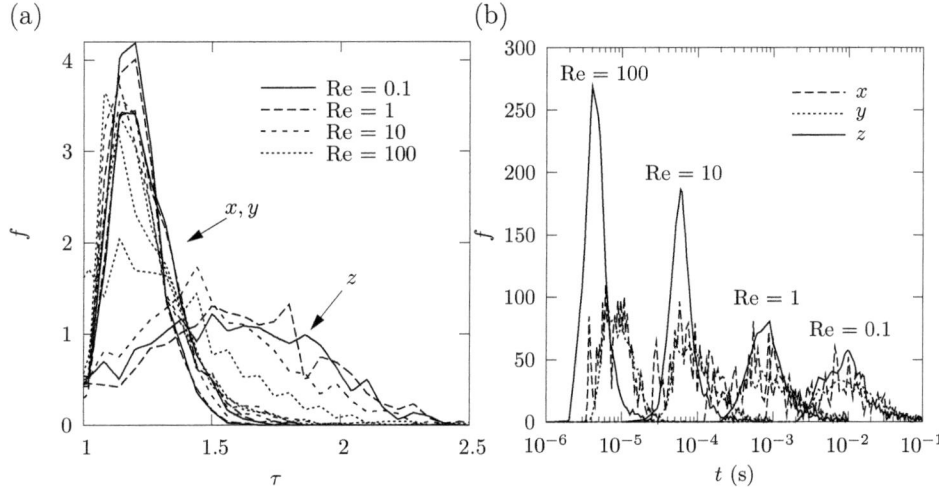

Figure 6.9: Tortuosity (a) and residence time (b) distributions for Re = 0.1 to 100 and for the three directions.

erated morphologies that are input to the DPLS methodology. Morphologies composed of bimodal distributed overlapping transparent sphere (BDOTS) in an opaque ceria matrix are examined, as these allow for a structured investigation of the morphology-property relation and for an adjustment of the effective transport properties to the specific process needs. BDOTS are employed because they closely resemble the ceria foams fabricated by the sacrificial pore-former process with two different pore-former graphite particle types (Alfa Aesar 10129 and 40769) with two distinct mean particle sizes (see figure 6.2.c). The porosity

Table 6.6: Mean, mode, and median tortuosity and residence time for sample no. 1 in the three directions for a sample with length 0.17mm at Re = 1.

Direction	τ_m	τ_{mode}	τ_{median}	t_m (s)	t_{mode} (s)	t_{median} (s)
x	1.24	1.20	1.22	0.0026	0.0006	0.0018
y	1.20	1.20	1.19	0.0028	0.0007	0.0020
z	1.61	1.80	1.59	0.0018	0.0009	0.0011

of BDOTS is calculated by

$$\varepsilon_{\text{BDOTS}} = 1 - \exp\left(-\frac{4\pi}{3}n\left[\xi r_1^3 + (1-\xi)r_2^3\right]\right), \tag{6.8}$$

where n represents the number density of pores, ξ represents the number fraction of larger pores, and $1-\xi$ the number fraction of smaller pores. r_i represents the mean radius of the pores. A 2D slice through the samples is shown in figure 6.10.

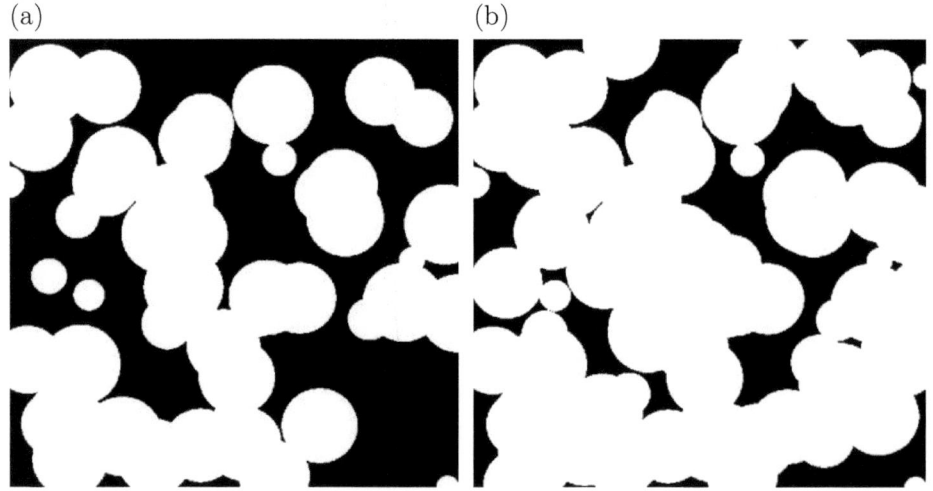

Figure 6.10: 2D slice trough the artificial BDOTS samples with $\xi = 1$, $r_1 = 50$ μm, $\varepsilon_{\text{BDOTS}} = 0.6$ (a) and 0.8 (b). Edge length is 540 μm.

Nu, K, and F_{DP} are calculated for samples made of $\xi = 1$, $\varepsilon_{\text{BDOTS}} = 0.6$ and 0.8, and $r_1 = 50$ μm. Figure 6.11.a shows the normalized pressure gradient over the porous sample. The resulting K and F_{DF} are given in table 6.7. K increases considerably (by more than an order of magnitude) when the sample porosity is increased, which is consistent with the hydraulic radius and fibrous

6.3. Tailored foam design

Table 6.7: Permeability, Dupuit-Forchheimer coefficient, and Nu correlation for the artificial porous sample with $\xi = 1$, and $\varepsilon_{BDOTS} = 0.6$ and 0.8.

ε_{BDOTS}	K (m^2)	F_{DF} (m^{-1})	Nu
0.6	$6.62 \cdot 10^{-11}$	219143	$3.03 + 0.55 Re^{0.71} Pr^{0.54}$
0.8	$2.13 \cdot 10^{-10}$	23315	$3.59 + 0.28 Re^{0.76} Pr^{0.55}$

bed models [102, 113]. Note that the capillary model, which only linearly relates K and ε, underestimates this dependence. In contrast, F_{DF} decreases with increasing porosity, which is a result of less tortuous paths for fluid flow across the sample, reducing the inertia-induced forces. This is consistent with F_{DF} being proportional to $(1-\varepsilon)/\varepsilon^3$ as predicted in [126]. Nu numbers are shown in figure 6.11.b. For low Re numbers (Re < 20 for Pr = 0.1, and Re < 3 for Pr = 1), heat transfer is favored for samples with high porosity. The opposite is true for high Re numbers, because vortices evolve in partially open pores (cavities) and obstruct the passage of fluid flow.

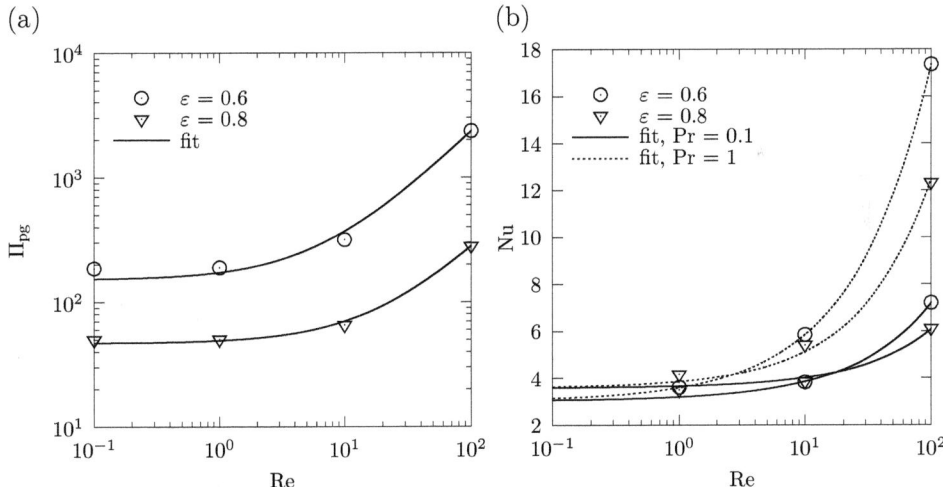

Figure 6.11: Normalized pressure gradient as a function of Re (a), and Nu numbers as a function of Re for Pr = 0.1 and 1 (b), for the artificially generated porous sample ($\xi = 1$ and $\varepsilon_{BDOTS} = 0.6$ and 0.8).

Comparison of the three ceria samples to the results of the BDOTS at $\varepsilon_{BDOTS} = 0.6$ shows that Nu numbers are considerably lower in the real samples. This is expected as r_1 (= 50 μm) is larger than the calculated mean diameters of the

ceria samples. Correspondingly, K is larger for the BDOTS.
The complete variation of the $\varepsilon_{\text{BDOTS}} - r_1 - r_2 - \xi$-set and the calculation of the corresponding effective transport properties enable in-depth understanding of the morphology-property relations and, consequently, guide the pore-level engineering.

6.4 Conclusions

Ceramic foams of ceria, used in redox reactions for the solar thermochemical splitting of H_2O and CO_2, were analyzed for their morphological characteristics and their effective transport properties. The samples were produced by a sacrificial pore-former process, using graphite particles with diameters in two different size ranges. The anisotropy of the primary particles and the uniaxial pressing during preparation result in structural anisotropy.

The ceramic foam's exact 3D micro-geometries were obtained by high-resolution tomography and were incorporated in direct pore-level numerical simulations to determine anisotropic effective heat transfer properties, i.e. extinction coefficients, thermal conductivities, and heat transfer coefficients as well as mass transfer properties, i.e. permeabilities, Dupuit-Forchheimer coefficients, tortuosity and residence time distributions, along the principal directions. Due to uniaxial pressing (z-direction), channels evolved along the x- and y-directions, which resulted in increased radiative extinction along the z-direction. The thermal conductivity along the x- and y-directions was adequately described by the parallel slab model. Convective heat exchange along the z-direction was enhanced, which can be explained with larger tortuosity. Calculated Nu values were within those predicted for packed beds. Permeability was higher, while the Dupuit-Forchheimer coefficient was lower in the z-direction compared to those in the other directions, and showed reasonable agreement with those calculated by capillary, hydraulic radius, fibrous bed and extended Ergun equation models.

Artificial porous ceria samples composed of bimodal distributed overlapping transparent spheres in an opaque medium - resembling the morphology of the ceria foam - were examined for enhanced transport properties. Permeability increased while Dupuit-Forchheimer coefficient decreased with increasing porosity. For large Re numbers, lower porosity resulted in enhanced convective heat transfer, while the opposite was true for high Re numbers because vortices evolved in partially open pores and obstructed the passage of fluid flow. The morphology-property relations and guidelines for pore-level engineering were elucidated.

The calculated effective properties can be incorporated volume-averaged mod-

6.4. Conclusions

els for accurate modeling and subsequent optimization of the solar reactor configuration [61]. The preliminary study on tailored foam design adjusted to the specific process needs can be extended and allows for foam engineering and consequently enhanced process performance.

Chapter 7

Reacting packed bed of carbonaceous material[1,2]

In this chapter, a reacting packed bed undergoing a high-temperature solid-gas thermochemical transformation is characterized for morphological and effective transport properties. The model reaction investigated is the solar gasification of carbonaceous material to syngas in a packed-bed reactor, which is introduced in the first part of this chapter. An analysis on using carbonaceous feedstock in different processes for the production of electricity is conducted to compare the energy efficiency and emission mitigation potential of the different processes involved.

CT is employed to obtain the exact 3D digital geometrical representation of the packed bed composed of complex, porous, and nonspherical particles, the morphology of which varies with time and process parameters (e.g., temperature, gasifying agent, partial pressure, and feedstock size) as the reaction progresses. This stands in contrast to previously investigated inert reticulate porous ceramic and anisotropic ceramic foams. Additionally, a packed bed is classified as a non-consolidated porous material, where contact resistance influences the effective transport properties. The morphological and effective heat and mass transfer properties are determined based on the CT data in the second part of this chapter. Experimental validation of the calculated morphological and effective radiative properties are reported.

Exemplary processes investigated by dynamic (unsteady) CT-based methodology are cement hydration [86], drying in granular media [109], bone ingrowth

---

[1]Material from this chapter has been published in: S. Haussener, W. Lipiński, P. Wyss, and A. Steinfeld. Tomography-based analysis of radiative transfer in reacting packed beds undergoing a solid-gas thermochemical transformation. *Journal of Heat Transfer*, 132: 061201, 2010. [83]

[2]Material from this chapter has been published in: S. Haussener, I. Jerjen, P. Wyss, and A. Steinfeld. Tomography-based determination of effective transport properties for reacting porous media. *Journal of Heat Transfer*, 134: 012601, 2012. [81]

into tissue-engineered scaffold materials [96], snow metamorphism under alternating temperature gradients [170], and foam growth [142].

The obtained effective properties can then be applied for the accurate derivation of the reaction kinetics and for the design and optimization of a packed-bed solar reactor [166].

7.1 Production of syngas by gasification of carbonaceous material

Gasification of carbonaceous feedstock is an chemical process occurring via pyrolysis, the thermal decomposition and devolatilization of the feedstock, and subsequent heterogeneous solid-gas reaction of the pyrolysis residue (char) with reactive gases (usually H_2O and/or CO_2). The desired product is syngas. A wide variety of carbonaceous feedstock (generally described by $CH_xO_yS_zN_u$) such as coal, biomass, agricultural, municipal and industrial waste, can be used. The process allows converting low-value or waste products to higher value products. If solid waste material, normally disposed by land filling, is used as feedstock, the required sites for waste disposal can be reduced.

Some of the most important factors of the syngas quality are feedstock composition, preparation and particle size, residence time, reactor heating rate, and reactor type (feed system, feedstock-reactant flow geometry, heat generation and transfer mode, and syngas clean-up system).

Steam and dry (CO_2) gasification of carbonaceous material can be described by the simplified net reactions:

$$CH_xO_yS_zN_u + (1-y)H_2O = \frac{1}{2}(x+2(1-y)-2z)H_2 + CO + zH_2S + \frac{1}{2}uN_2, \quad (7.1)$$

$$CH_xO_yS_zN_u + (1-y)CO_2 = \frac{1}{2}(x-2z)H_2 + (2-y)CO + zH_2S + \frac{1}{2}uN_2. \quad (7.2)$$

In reality, several concurrent reactions take place, especially in the low temperature region [176]. Combinations of steam and dry gasification are possible and allow for the production of the desired syngas quality.

The model feedstock used are waste tire shreds, which are nowadays mostly disposed by land filling. Worldwide $6 \cdot 10^6$ t/y (EU: $1.5 \cdot 10^6$ t/y, Switzerland: $0.05 \cdot 10^6$ t/y) of waste tire have to be disposed. Generally, waste disposal (which is achieved by land filling up to 60% today) could be avoided or at least reduced by waste-to-energy solutions, such as gasification [4].

7.1. Production of syngas by gasification of carbonaceous material

The equilibrium composition, calculated with Gibbs' free enthalpy minimization [157], for the stoichiometric systems of steam and dry gasification of waste tire shreds ($x = 1.035$, $y = 0.029$, $z = 0.004$, $u = 0.008$) at 0.1, 1 and 10 bar is shown in figure 7.1. Species with fractions below 1% are not shown. Higher pres-

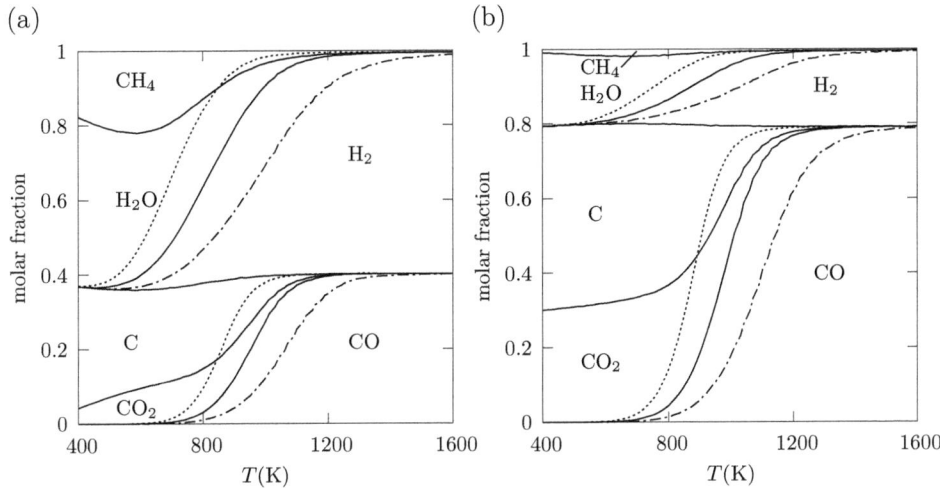

Figure 7.1: Equilibrium composition of stoichiometric steam gasification (a) and dry gasification (b) for 0.1 bar (dash-dotted line), 1 bar (solid line), and 10 bar (dotted line).

sure shifts the equilibrium position to the left, due to Le Chatelier's principle, which states that higher pressure changes the position of the equilibrium in such a way that the pressure change is counteracted. The steam gasification at 1 bar goes to completion at 1300 K while the CO_2 gasification needs 1500 K. Lower pressures lower the needed reaction temperature considerably (e.g., steam gasification at 0.1 bar needs 1100 K for completion). Over- and understoichiometric equilibrium compositions of reactions (7.1) and (7.2) are shown in figure 7.2. Understoichiometric reactions leads to unreacted carbon while overstoichiometric reaction leads to excess of the gasifying agent and increased production of CO_2.

Steam and dry gasification of waste tire shreds ($LHV_{wts} \approx 38$ MJ/kg) to syngas has the potential to increase the low heating value (LHV) by 23 and 20%, respectively.

An energy analysis is carried out in order to examine the performance of waste tire shreds used as fuel for the production of electricity. Five different

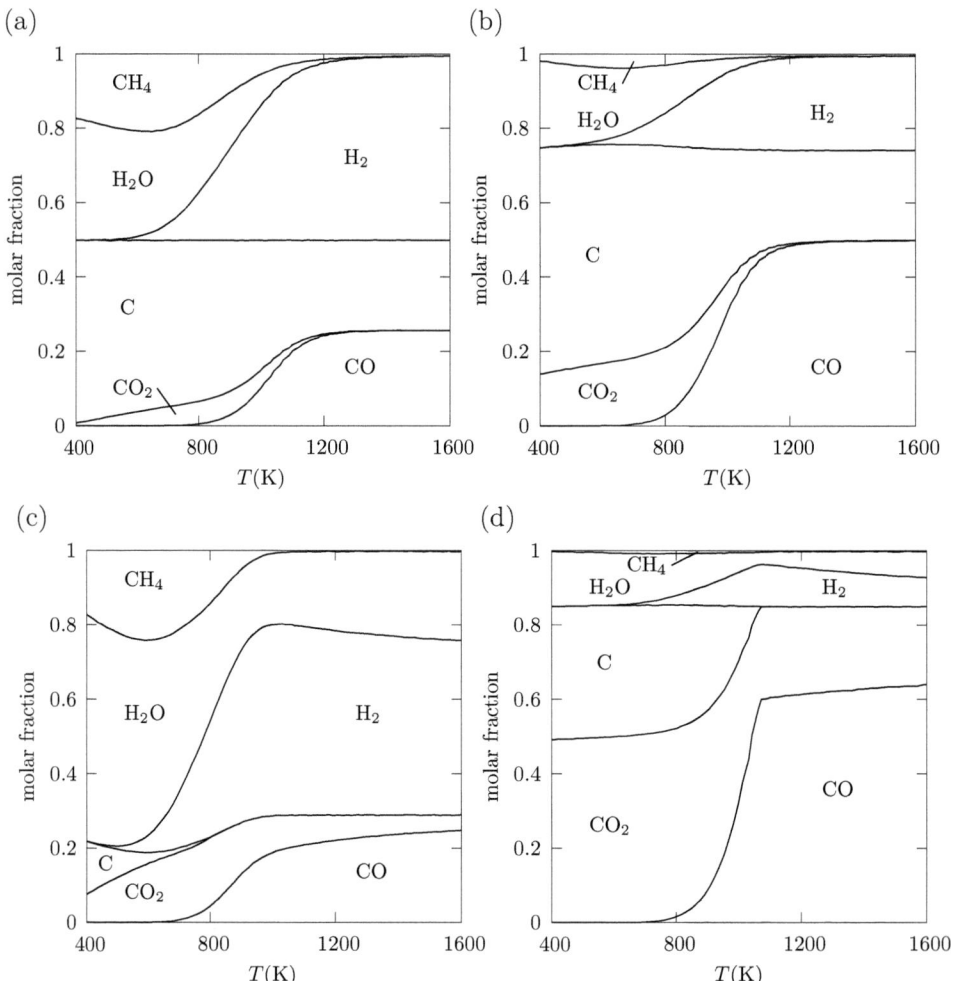

Figure 7.2: Equilibrium composition at 1 bar of steam (a,c) and dry (b,d) gasification for understoichiometric (stoichiometry factor = 0.5) (a,b) and overstoichiometric (stoichiometry factor = 2) (c,d) reactions.

7.1. Production of syngas by gasification of carbonaceous material

systems are investigated:

- Tires are used to fuel a Rankine cycle (RC)
- Tires are solar gasified and the syngas is used to fuel a combined Brayton-Rankine cycle (SG+CC)
- Tires are solar gasified, the syngas is processed to hydrogen by water-gas sift reaction and used to fuel a H_2/O_2 fuel cell (SG+WGS+FC)
- Tires are solar gasified, the syngas is used in a high-temperature solid-oxide fuel cell, its waste heat is used to drive a Rankine cycle (SG+FC+RC)
- Tire shreds are autothermally gasified and the syngas is used to fuel a combined Brayton-Rankine cycle (AG+CC)

Waste tire shreds feedstock (fs) is compared to using anthracite as feedstock. The analysis follows partially the analysis described in [223]. The baseline parameters and assumptions for the five systems are: feedstock mass flow of 1 g/s desulfurized and cleaned carbonaceous material, ambient pressure, Rankine cycle with 35% efficiency, solar reactor of black-body type at 1350 K and radiation input of 2000 kW/m² at its aperture, combined cycle with 55% efficiency, water-gas shift reaction processes exothermally at 700 K, separator based on pressure swing adsorption with recovery rate of 90% and its energy expenditure is taken from the energy produced by the fuel cell, H_2/O_2 fuel cell with 65% efficiency, solid-oxide fuel cell with 65% efficiency processes at 1300 K, and ideal autothermal gasification (gasification reaction enthalpy matches partial oxidation reaction enthalpy). The system efficiency,

$$\eta = \frac{\dot{W}}{q_{\text{sol}} + \dot{m}_{\text{fs}} \text{LHV}_{\text{fs}}}, \tag{7.3}$$

specfic electric output,

$$eo = \frac{\dot{W}}{\dot{m}_{\text{fs}}}, \tag{7.4}$$

electric gain factor,

$$egf = \frac{\dot{W}}{\dot{W}_{\text{RC}}}, \tag{7.5}$$

and the specific CO_2 emissions,

$$em = \frac{(M_\text{C} + 2M_\text{O})\dot{n}_{\text{CO}_2}}{\dot{W}}, \tag{7.6}$$

Table 7.1: Efficiency, specific electrical output, electrical gain factor, and the specific CO_2 emissions of the five different electricity generation routes using tire waste shreds or anthracite as feedstock and H_2O or CO_2 as gasifying agent.

		η	em (kg_{CO_2}/kWh)	egf	eo (kWh/kg_{fs})
RC	tire	0.35	0.885	1	3.68
	anthracite	0.35	0.985	1	3.47
		H_2O gasification			
SG+CC	tire	0.44	0.448	1.98	7.28
	anthracite	0.44	0.506	1.95	6.76
SG+WGS+FC	tire	0.53	0.384	2.32	8.48
	anthracite	0.51	0.438	2.27	7.80
SG+FC+RC	tire	0.64	0.314	2.82	10.37
	anthracite	0.63	0.355	2.77	9.61
AG+CC	tire	0.45	0.689	1.28	4.73
	anthracite	0.44	0.778	1.27	4.39
		CO_2 gasification			
SG+CC	tire	0.47	0.832	2.10	7.73
	anthracite	0.48	0.932	2.08	7.23
SG+WGS+FC	tire	0.52	0.761	2.32	8.44
	anthracite	0.51	0.868	2.27	7.76
SG+FC+RC	tire	0.67	0.592	2.95	9.38
	anthracite	0.66	0.665	1.91	8.77
AG+CC	tire	0.48	0.649	1.36	5.02
	anthracite	0.47	0.727	1.35	4.70

for the five systems are given in table 7.1. Gasifying agents analysed are H_2O or CO_2.

Systems which use solar energy to pre-process the products result in system efficiencies above 0.44. This is considerably larger than in the conventional RC with an efficiency of 0.35. The egf nearly doubles for the solar routes and the specific CO_2 emissions are reduced by more than a factor two. Autothermal gasification only allows for an egf of approximately 1.3 and somewhat lower reduction in CO_2 compared to the solar routes. Waste tire shreds and anthracite perform equally well. Quenching of the product gases at different steps in the five systems investigated allows for heat recovery and increases the system per-

7.2. Experimental campaign for sample production

formance.

Three directly or indirectly irradiated reactor concepts for the solar gasification of carbonaceous material have been proposed: packed-bed reactors [69, 70, 166], fluidized-bed reactors [94, 145, 55, 107, 224, 106], and particle-flow/aerosol reactors [242, 137, 117]. Fluidized-bed and particle-flow/aerosol reactors exhibit enhanced heat and mass transfer properties and consequently higher efficiencies. Packed-bed reactor on the other side convince by their relatively high-contacting area, and by their simple design and operation. The indirectly irradiated solar reactor used for coal gasification in [166] is shown in figure 7.3. The carbonaceous material builds a packed bed in the reaction chamber. It absorbs the radiation from the emitting plate of the upper chamber and serves as reactant and reaction side for the gasification reaction.

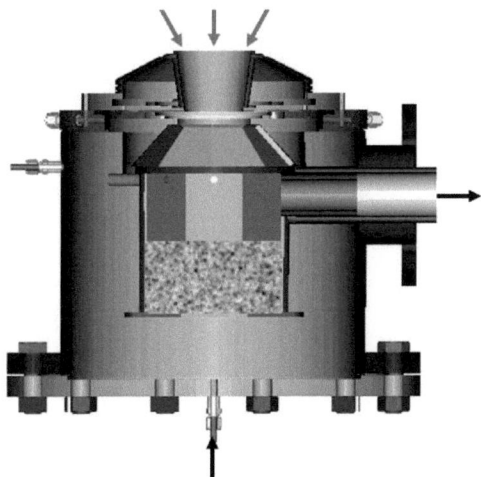

Figure 7.3: Indirectly irradiated solar reactor used for coal gasification [166], hosting a packed bed of carbonaceous material in the reaction chamber.

7.2 Experimental campaign for sample production[3]

A packed bed of tire shreds is selected as the model reactor. The gasification of this waste carbonaceous material into high-quality synthesis gas is investigated

---
[3]Material from this chapter is partially based on work performed in the framework of: J. Gaabab, Experimental investigation of the morphological changes in a packed bed of tire shreds undergoing gasification, Semester thesis, ETH Zurich, 2008. [62]

in a packed-bed reactor using concentrated solar energy as the source of high-temperature process heat [167]. In this study, a laboratory packed-bed reactor, schematically shown in figure 7.4, is used to conduct the gasification reaction at controlled conditions and to produce sample materials at different reaction extents. Prior to the actual gasification reaction pyrolysis takes place. These

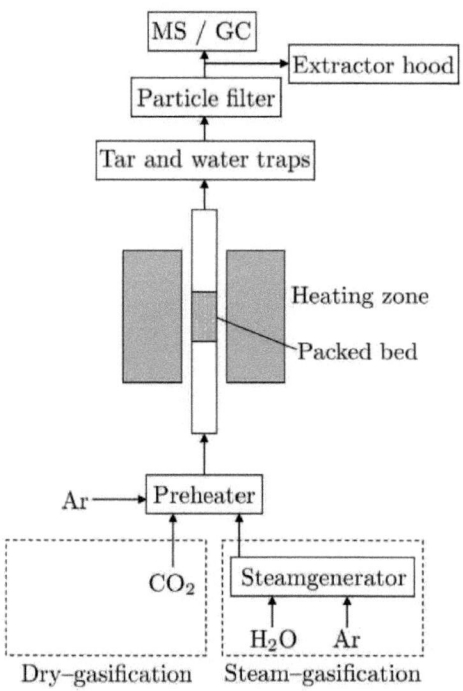

Figure 7.4: Schematic of the tubular packed-bed reactor setup used for the gasification of carbonaceous materials.

samples are then scanned by tomography. Their BET specific surface area is measured by N_2 adsorption (Micromeritics TriStar 3000) and their particle-size distribution is measured by laser scattering (HORIBA LA-950 analyzer). Outlet gas composition during gasification is monitored by mass spectrometry (MS, Pfeiffer Vacuum OmniStar GSD 301 O1) and gas chromatography (GC, Varian CP-4900 Micro GC). Proximity analysis of the tire shreds indicates 63 wt% volatiles, 29 wt% fixed carbon, 7 wt% ash and 1 wt% moisture. Elemental analysis indicates 82 wt% C, 7 wt% H, 3 wt% O, 2 wt% S, and heavy metal impurities. Energy dispersive X-ray spectrometer analysis shows that the main

7.2. Experimental campaign for sample production

components in the ash are Si, Zn, and Fe based oxides.

Samples are first pyrolyzed at 1000 K (heating rate approximately 90 K/min) to release volatiles. Approximately 3 g of pyrolyzed material is loaded in a 2.6 cm inside diameter quartz tube, rapidly heated by a radiative source, and gasified either by steam or CO_2 diluted by Ar. Once the reaction reaches a desired carbon conversion, the reaction is stopped (stop hating and reacting gas inflow), the quartz tube is removed from the furnace and rapidly cooled (rate approx. 70 K/min) in an oxygen free atmosphere. Therefore, no significant changes in morphology due to cooling are expected. The partially reacted sample is extracted. The carbon conversion (or reaction extent) is defined by

$$X_C = 1 - \frac{n_C}{n_{C,0}} = \frac{\int_0^1 \dot{n}_{Ar} x_{Ar}^{-1} (x_{CO_2} + x_{CO} + x_{CH_4}) \, dt}{\int_0^\infty \dot{n}_{Ar} x_{Ar}^{-1} (x_{CO_2} + x_{CO} + x_{CH_4}) \, dt}. \qquad (7.7)$$

The process parameters are listed in table 7.2: furnace temperature, type of gasifying agent, partial pressure of gasifying agent, and type of feedstock. Carbon conversions, starting after pyrolysis ($X_C = 0.68$), are shown in figure 7.5 as a function of reaction time for the five different experimental runs. Samples at $X_C = 0$ (initial), 0.68 (after pyrolysis, called char), 0.79, 0.9, and 1 (ash) obtained in each of the five experiments are selected for the subsequent analysis.

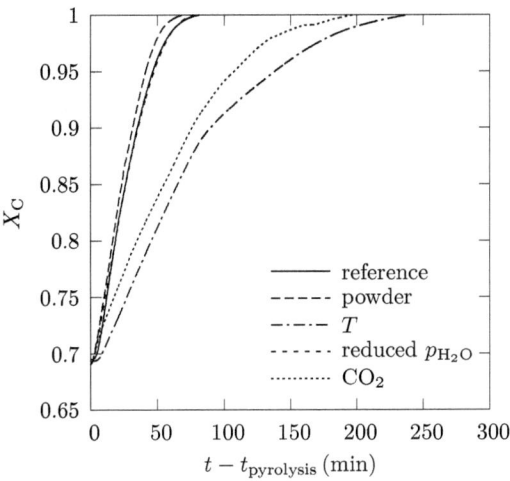

Figure 7.5: Carbon extent X_C, starting after pyrolysis, as a function of reaction time for five different sets of process parameters, as described in table 7.2.

Table 7.2: Process parameters for the five experimental runs.

Case	T (K)	Gasifying agent	p_{ga} (bar)	Feedstock (mm)
Reference	1273	H_2O	0.8	Granular, $d_m = 1$
Powder	1273	H_2O	0.8	Powder, $d_m = 0.5$
T	1173	H_2O	0.8	Granular, $d_m = 1$
Reduced p_{H_2O}	1273	H_2O	0.4	Granular, $d_m = 1$
CO_2	1273	CO_2	0.8	Granular, $d_m = 1$

7.3 Morphological characterization

7.3.1 Computed tomography

LRCT of the samples are obtained by exposing them to an unfiltered X-ray beam generated by electrons incident on a wolfram target. The generator is operated at 40–50 keV and a current of 0.2–0.3 mA. A Hamamatsu flat panel C7942 CA-02 protected by a paper filter is used to detect the transmitted X-rays. The samples are scanned at 1800 angles (projections). Each projection consists of an average of six scans at 1.2 s exposure time. The scans are performed for voxel sizes of 10 µm (at $X_C = 0, 0.68, 0.79, 0.9$) and 5 µm (at $X_C = 1$). The fields of view investigated are 11.2 x 11.2 x 12 mm³ and 5.6 x 5.6 x 6 mm³, respectively.

In addition, HRCT (voxel size of 0.37 and 3.7 µm and field of view 0.76 x 0.76 x 0.62 mm³ and 7.6 x 7.6 x 7.6 mm³, respectively) are obtained with synchrotron radiation at the TOMCAT beamline at SLS of PSI [208, 200]. The scans are obtained for 10 and 23 keV photon energy, respectively, 0.4 mA beam current, 1.5 and 0.94 s exposure time, respectively, and 1500 projections. Figure 7.6 shows tomograms obtained with HRCT (voxel size of 3.7 µm) for the reference case sample at $X_C = 0$, 0.68 and 1 (figure 7.6.a, b and, c) and with submicron HRCT (voxel size of 0.37 µm) for a single particle of the reference case sample at $X_C = 0.79$ (figure 7.6.d). The latter scan was performed to examine the amount of pores below 3.7 µm. Table 7.3 summarizes the scans with different resolution obtained for the samples.

The data obtained by tomography is digitally processed by brightness and contrast adjustment, and by intensity transformation, obtained via a two-step gamma correction. When appropriate, median filtering is applied to reduce the noise. Due to the highly heterogeneous material containing optically thin carbon-containing compounds and optically thick heavy metal impurities, phase segmentation is complicated. Local multistep threshold segmentation, imple-

7.3. Morphological characterization

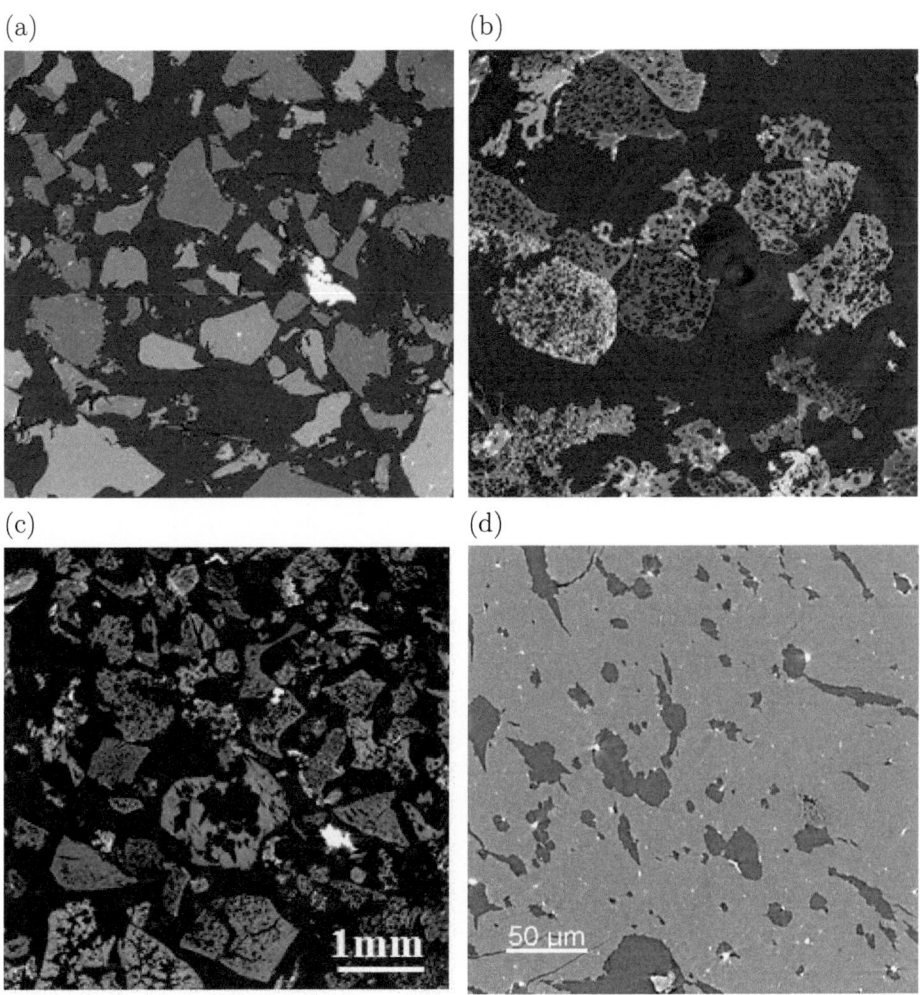

Figure 7.6: Tomograms of the reference case sample with voxel size of 3.7 µm: (a) initially at $X_C = 0$, (b) after pyrolysis at $X_C = 0.68$ (char), and (c) after gasification at $X_C = 1$ (ash). The scale bar in (c) applies to (a) and (b). In (d) a high-resolution tomogram (voxel size of 0.37 µm) of a single carbon particle at $X_C = 0.79$ is shown.

Table 7.3: Resolution of tomographic scans obtained for the packed bed of carbonaceous material at different X_C.

Sample	Low–resolution		High–resolution at SLS	
	10 μm voxel	5 μm voxel	3.7 μm voxel	0.37 μm voxel
$X_C = 0$	x		x	x
$X_C = 0.68$	x		x	x
$X_C = 0.79$	x		x	x
$X_C = 0.9$	x		x	x
$X_C = 1$		x	x	x

mented in MATLAB, is used to allow for more accurate phase detection. Adjustment of calculated and experimentally determined porosity is not used for threshold selection as it results in wrong phase segmentation due to insufficient resolution of the nanopores, which account for 10–20% of the porosity. The iso-surface describing the phase interface is obtained when the continuous density value equals the threshold value for phase segmentation. A representative rendered 3D geometry is shown in figure 7.7.

Figure 7.7: A 3D rendered geometry of the reference case at $X_C = 0.68$ with cube length of 1.5 cm.

7.3.2 Porosity and specific surface area

Experimental determination

Porosity is determined by weight measurements, when the approximate intrinsic density is assumed to increase linearly with decreasing carbon content, and is given by $\rho = \rho_{ash} X_C + \rho_C (1 - X_C)$ with $\rho_C = 1700$ kgm^{-3} and $\rho_{ash} = 2500$ kgm^{-3}. ρ_{tire} of the initial tire shreds (before pyrolysis) is measured to be 1200 kgm^{-3} by He pycnometry (AccuPyc 1330). In figure 7.8, ε during gasification is shown as a function of the carbon conversion for the five experimental runs listed in table 7.2. The porosity peaks at $X_C = 0.86$ as a result of growing

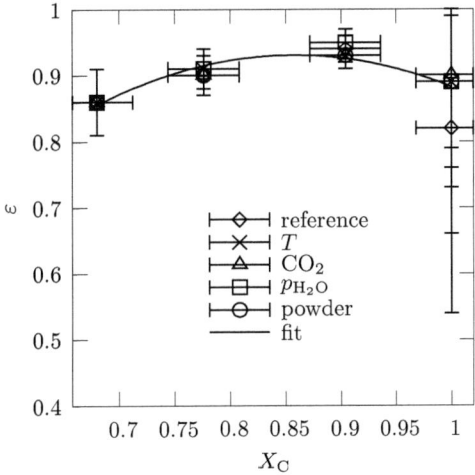

Figure 7.8: Experimentally determined porosity for the packed bed of tire shred as a function of carbon conversion for the five experimental runs listed in table 7.2.

pores and break-up of fragile particles during gasificaiton [125]. The measured values correspond to a loosely packed bed of randomly oriented and located nonspherical particles having uniform size and sphericity (fraction of surface area of volume–equivalent sphere to surface area of particle) smaller than 0.25 [235], indicating highly porous particles. The measured porosity during gasification is fitted to a second order polynomial function (RMS = 0.002)

$$\varepsilon_{ex}(X_C) = -2.208 X_C^2 + 3.789 X_C - 0.699. \quad (7.8)$$

The porosity of the unreacted packed bed of tire shreds is determined to be $\varepsilon_{ex} = 0.60 \pm 0.05$, which corresponds to a packed bed of nonspherical particles of uniform size and with sphericity of 0.55 [235]. If this value is included in the fit, one obtains (RMS = 0.002)

$$\varepsilon_{\text{ex}}(X_C) = -0.366 X_C^2 + 0.673 X_C + 0.593. \tag{7.9}$$

The BET specific surface area and the corresponding fraction resulting from nanopores ($d_{\text{pore}} < 2$ nm) is shown figure 7.9 as a function of the carbon conversion for the five experimental runs listed in table 7.2. Before pyrolysis, BET surface area is 0.6 m^{2}g^{-1} for the granular and 1.6 m^{2}g^{-1} for the powder feedstock, and no nanopores are detected. During pyrolysis, it increases to 70 m^{2}g^{-1} of which a small fraction ($\sim$5%) is associated to nanopores. During gasification, the BET specific surface area increases up to $\sim$ 700 m^{2}g^{-1} for $X_C = 0.9$ and decreases for the residual ash ($X_C = 1$), which is consistent with the variation in porosity. The fraction of nanopores increases and peaks at 60% for $X_C = 0.79$. No nanopores are detected in the ash. The different values obtained for H_2O and CO_2 gasifying agents are presumably the result of different mechanisms as CO_2 mainly reacts at the external surface while H_2O diffuses to the particle core [40].

In general, the variation in the reaction temperature, partial pressure, gasifying agent, and particle-size (as described in table 7.2) do not significantly affect the morphology of the sample at the same carbon conversion.

Numerical determination

Two-point correlation function is used to determine porosity and specific surface area. The calculated porosity of the unreacted packed bed of 0.61 compares well to the experimentally determined one of 0.60 ± 0.05. Figure 7.10 shows the experimentally measured and numerically calculated porosity as a function of the carbon conversion during gasification for the reference case. The failure in predicting the porosity and its increase with increasing X_C is related to the resolution of the tomographic scans, which is limited by the tomographic setup, the subsequent image processing (especially filtering) and the relative increase in optically thick material, which distorts the tomographic image. The impact of the insufficient scanning resolution and subsequent image processing is roughly calculated to be $(1 - \varepsilon)\varepsilon_{\text{sub}} \approx 0.02$, where ε_{sub} (≈ 0.2 for $X_C = 0.79$) is the porosity of the particle only detectable by submicron high-resolution tomography. Nanopores are not detectable, but obviously present as is indicated by the BET measurements.

Calculated specific surface shows an increase up to $X_C = 0.9$ but the experimentally observed decrease for the ash cannot be elucidated in this study.

7.3. Morphological characterization

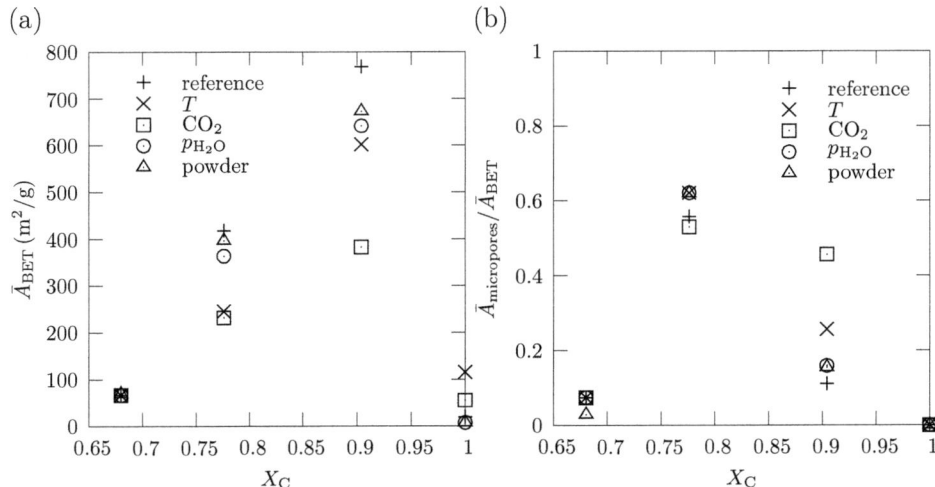

Figure 7.9: Experimentally determined specific surface area (a) and the corresponding fraction resulting from micropores (b) as a function of carbon conversion for the five experimental runs listed in table 7.2.

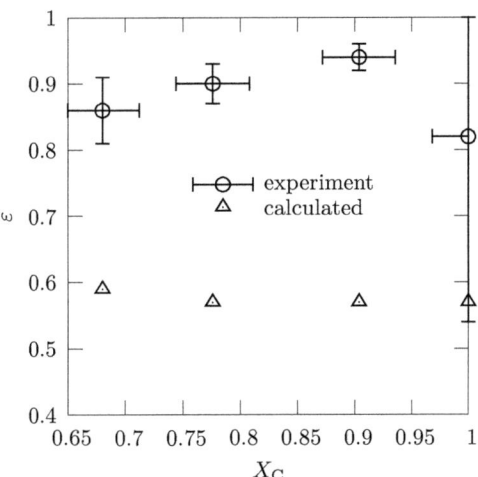

Figure 7.10: Experimentally measured and numerically calculated porosity as a function of the carbon conversion for the reference case.

7.3.3 Particle-size distribution

Experimental determination

The experimentally measured (volume-based) particle-size distribution is shown in figure 7.11.a for the reference case at $X_C = 0$, 0.68, 0.79, 0.9, and 1. As expected, the main peak shifts to the left as particles shrink, and the small peaks associated with smaller particles resulting from particle break-up increase during the reaction. Note that these distributions are qualitative as particles are not spherical. Mean and median diameters of the measured distributions are given in table 7.4.

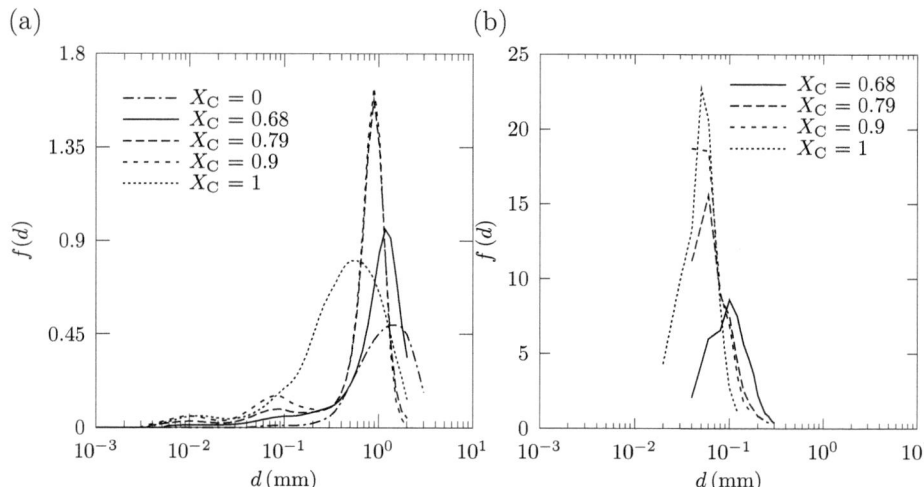

Figure 7.11: Experimentally measured (a) and numerically calculated (b) size distribution of the particles for the reference case at various carbon conversions $X_C = 0$, 0.68, 0.79, 0.9, and 1.

Numerical determination

The numerically calculated particle-size distributions, shown in figure 7.11.b for the reference case at $X_C = 0.68$, 0.79, 0.9, and 1, are based on the largest sphere that fits inside the particle. Therefore, for nonspherical, complex, porous and fractal-like particles, as evolving during pyrolysis, these distributions deviate from those experimentally measured. The calculations are limited by the voxel size of the CT scans ($d_{\min} = 4 \cdot$voxel size). Since the particle-size distribution is calculated based on the solid phase, the limited resolution of the CT scans leads

7.4. Heat transfer characterization

Table 7.4: Mean and median diameter of the particles in the packed bed as function of reaction extent.

Sample at X_C	$d_{m,ex}$ (mm)	$d_{median,ex}$ (mm)
0	1.21	1.14
0.68	0.99	0.97
0.79	0.71	0.75
0.9	0.68	0.73
1	0.54	0.42

to an over-prediction of the particle-size due to virtual particle agglomeration. The measurements (see figure 7.11.a) show that the amount of particles in the 10 µm range is small compared with the one in the 100 µm range. Therefore the influence of this distortion is assumed negligible on the particle-size distribution in the range of 60 to 100 µm. Indeed, an increase in small particles due to shrinkage and break-up of the initial particles during the reaction is observed in the numerically calculated size distributions.

7.3.4 Representative elementary volume

REV is determined by calculating the porosity of a subsequently growing subsample until its variation is within a tolerance band of ±0.05. The edge length of the REV, l_{REV}, was found to be 5 mm, independent of the process parameters.

7.4 Heat transfer characterization

7.4.1 Radiative characterization

The packed bed of the carbonaceous material is assumed to be opaque for visible and near-IR radiation, which is the spectral range encountered in the solar-driven reactor [167]. The fluid phase is assumed to be radiatively nonparticipating. Hence, the variation in the radiative intensity in continuum models is described by a single equation of radiative transfer as derived in chapter 2. LRCT data is used for the calculation of the radiative properties. Since the smallest pores or particles detected by CT and consequently employed in the analysis are larger than the voxel sizes of the scans, geometric optics can be assumed for radiation wavelengths smaller than 1 µm [141].

The collision-based MC method is applied to compute the cumulative distribution functions of the radiation mean free path and of the cosine of incidence at the solid wall, and consequently extinction coefficient, scattering coefficient and scattering phase function of the packed bed of waste tire shreds at different reaction extents, see chapter 3.

The absorption characteristics of the samples and the contribution of dependent scattering vary with the reaction extent since ash is less absorbing than coal ($\rho_{r,sp,C} = 0.273$, $\rho_{r,d,C} = 0.1$, $\rho_{r,sp,ash} = 0.092$, and $\rho_{r,d,ash} = 0.75$ [216, 96]). Gas, packed-sphere, liquid, and modified liquid models are used to estimate the deviations of the scattering and absorption coefficients from the corresponding values obtained by assuming independent scattering [214]. For a packed bed with $d_m = 0.4$ and 1 mm (d_m of particle-size distribution shown in figure 7.11.a at $X_C = 1$ and 0.68, respectively), the maximum deviation of the scattering efficiency (appearing at the largest radiation wavelength in our spectral range of interest 1 µm is 5% and 23% for $\varepsilon_{ex} = 0.82$ and 0.86, respectively (measured and depicted in figure 7.8 at $X_C = 1$ and 0.68, respectively). Therefore, dependent scattering effects are neglected in the radiative transfer analysis.

The cumulative distribution function of the cosine of incidence at the solid wall and the scattering phase function are computed for two limiting cases: a specular and a diffuse solid-gas interface. For tire shreds, a combination of these two cases is anticipated to be valid. The specular directional-hemispherical reflectivity is calculated using Fresnel's equations for the complex refractive index of the carbon-ash mixture $m = (1 - X_C)m_C + X_C m_{ash}$, where $m_C = 2.2 - 1.1i$ is the complex refractive index of carbon and $m_{ash} = 1.5 - 0.02i$ is the complex refractive index of ash [141, 184]. The refractive index of tire is estimated to be $m_{tire} = 1.38 - 1.12i$ (tires main components are styrene-butadiene-copolymer and carbon black). Note that the cumulative distribution function of radiation mean free path, and consequently, the extinction coefficient, are independent of the interface reflection type in the geometrical optics range.

The extinction coefficient β and scattering phase function Φ are shown in figure 7.12 for the reference case at $X_C = 0$, 0.68, 0.79, 0.9, and 1. β increases with X_C as particles shrink and shorten the attenuation path length. An empirical correlation of the extinction coefficient inversely proportional to the characteristic diameter supports this trend [87]. The extinction coefficient is fitted to an exponential function (RMS = 22.3):

$$\beta_{MC}(X_C) = 3968 + 0.00138 \exp(16.0 X_C). \tag{7.10}$$

The scattering phase function is independent of the reaction extent for the assumed diffusely reflecting interface. This result is consistent with the small

7.4. Heat transfer characterization

Table 7.5: Coefficients of the exponential fit to the scattering phase function for specularely reflecting solid-gas interface as a function of X_C.

X_C	a_1	a_2	a_3	RMS
0	0.804	8.430·10^{-2}	2.431	0.007
0.68	0.986	5.031·10^{-7}	13.55	0.017
0.79	0.899	1.415·10^{-2}	4.057	0.001
0.9	0.750	3.622·10^{-2}	4.017	0.009
1	0.614	6.377·10^{-3}	6.697	0.124

differences obtained between the phase functions for diffuse reflecting identical overlapping transparent spheres and for diffuse reflecting identical overlapping opaque spheres [211, 165, 82], although they differ largely in morphology. Φ is described by a second order polynomial function (RMS = 0.01):

$$\Phi_d = 0.565\mu_s^2 - 1.394\mu_s + 0.812. \tag{7.11}$$

In contrast, the scattering phase function for specularly reflecting particles exhibits a large forward scattering peak. This peak decreases during pyrolysis as the real part of the tire particle increases. During gasification the peak is enhanced with increasing X_C due to the decrease in the real part of the refractive index of the carbon-ash particle. The coefficients of the exponential fit,

$$\Phi_{sp} = a_1 + a_2 \exp(a_3\mu_s), \tag{7.12}$$

are listed in table 7.5. The scattering albedo (σ_s/β) during gasification ($X_C = 0.68$–1) can be approximately calculated as:

$$\frac{\sigma_s}{\beta} = (1 - X_C)\rho_{r,C} + X_C\rho_{r,ash}, \tag{7.13}$$

for the specular solid-gas interface, $\sigma_s/\beta = 0.273$ and 0.092 at $X_C = 0$ and 1, respectively. For the diffuse solid-gas interface $\sigma_s/\beta = 0.1$ and 0.75, respectively, [216, 96].

The extinction coefficient of the unreacted packed bed of tire shreds ($X_C = 0$) is experimentally estimated with the spectroscopy system that is described in chapter 5.2.3. $\beta_{ex} = 4976 \pm 1180$ m^{-1} compares well with the numerically determined extinction coefficient at $X_C = 0$ of $\beta = 4172$ m^{-1}.

7.4.2 Conductive characterization

The effective conductivity as a function of the fluid and solid conductivity (k_f/k_s) is determined for the reacting packed bed. The tomography data of the HR scans

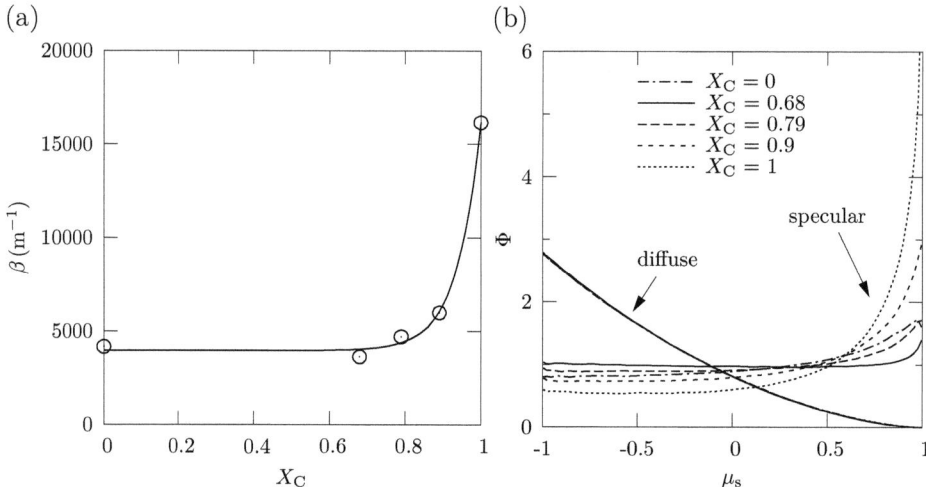

Figure 7.12: (a) Extinction coefficient as a function of carbon conversion (markers) and its fit (solid line) given by eq. (7.10), and (b) scattering phase function for the reference case at various carbon conversions $X_C = 0$, 0.68, 0.79, 0.9, and 1 for diffusely and specularly reflecting particles.

is used (voxel size 3.7 µm) for $X_C = 0$ and 0.68, and of the LR scans (voxel size 5 µm) for $X_C = 1$. Samples of 5.18 x 5.18 x 2.59 m³ and 4 x 4 x 4 mm³, respectively, limited by FOV of the scans and larger than REV, are used for the conduction calculations. Contact resistance between the particles is neglected. The normalized effective conductivity is shown in figure 7.13. It decreases as the reaction progresses due to increase in porosity (during pyrolysis) and particle shrinkage (during gasification), and lies between the values predicted by the parallel and serial slab models, describing maximum and minimum effective conductivities for a two-phase media [102].

7.4.3 Convective characterization

The tomography data of the HR scans is used (voxel size 3.7 µm) for $X_C = 0$ and 0.68, and of the LR scans (voxel size 5 µm) for $X_C = 1$. A sample with dimensions of 4.7 x 4.7 x 2.3 mm³ is used for the calculations, limited by the FOV of the CT scans. This volume corresponds to a cubic edge length of 3.7 mm. It results in a half bandwidth $\delta < 0.054$ when using the porosity-based definition of l_{REV}. ε calculated based on the absorption value in each voxel ($\varepsilon = N_{vox,void}/N_{vox,tot}$) are 0.61, 0.74 and 0.65 for $X_C = 0$, 0.68 and 1,

7.4. Heat transfer characterization

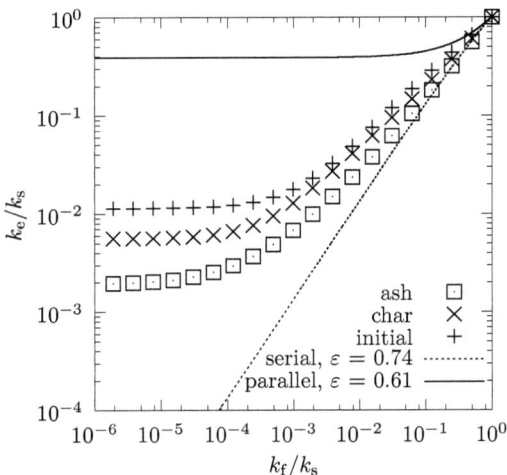

Figure 7.13: Effective conductivity (normalized by solid conductivity) of the packed bed initially, after gasification (char) and after full conversion (ash) in function of the fluid to solid conductivities. Normalized effective conductivity of serial and parallel slab models are shown for the numerically determined minimum and maximum porosities of the packed bed.

respectively. The Re number is based on the experimentally determined median diameter, $d_{\text{median,ex}}$, given in table 7.4.

Convergence of the numerical calculation is achieved for a terminal residual RMS of the iterative solution below 10^{-4} and for a maximal mesh element length of 90 µm (initial and char) and 117 µm (ash). The meshes are composed of approximately $3.5 \cdot 10^7$ tetrahedral elements. The meshes are generated by the in-house generator for unstructured body-fitted grids [57, 58]. Two quad-core Intel Xeon 2.5 GHz processors and 32 GB RAM are used to solve the governing equations in approximately 24 h. Alternatively, using twelve AMD Opteron 2.5 GHz processors and 42 GB RAM of ETH's high-performance cluster Brutus reduces computational time by a third.

The calculated and fitted Nu correlations for the packed bed at different X_C are given in table 7.6. Figure 7.14 shows the numerically calculated Nu numbers for the packed bed at Pr = 0.1, 1, and 10 and their fits. The four constants of the Nu correlation are related to the reaction extent by a 2nd-order polynomial function,

$$a_1(X_C) = 23.92 X_C^2 - 27.66 X_C + 8.34, \qquad (7.14)$$

Table 7.6: Fitted Nu correlations and RMS of the fit for the packed bed at X_C = 0, 0.68 and 1.

	Nu	RMS
Initial ($X_C = 0$)	$8.399 + 0.234 \text{Re}^{0.909} \text{Pr}^{0.627}$	3.928
Char ($X_C = 0.68$)	$0.754 + 0.091 \text{Re}^{0.740} \text{Pr}^{0.570}$	0.288
Ash ($X_C = 1$)	$4.592 + 0.274 \text{Re}^{0.886} \text{Pr}^{0.684}$	2.338

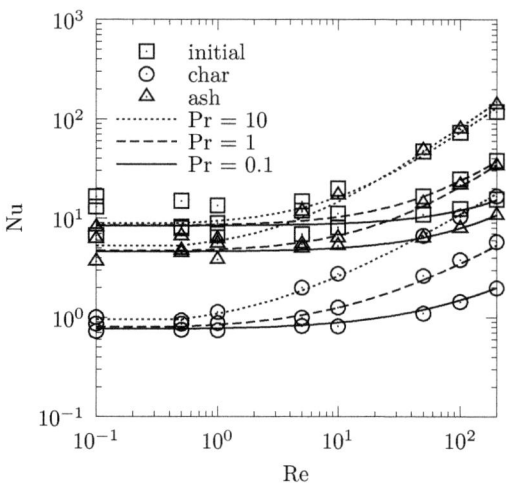

Figure 7.14: Re-dependent Nu numbers for the packed bed at X_C = 0, 0.68 and 1 (initial, char and ash) and at Pr = 0.1, 1, and 1. The dots indicate the numerically calculated Nu numbers and the lines indicate the fits given in table 7.6.

$$a_2(X_C) = 0.83 X_C^2 - 0.79 X_C + 0.23, \quad (7.15)$$

$$a_3(X_C) = 0.74 X_C^2 - 0.76 X_C + 0.91, \quad (7.16)$$

$$a_4(X_C) = 0.47 X_C^2 - 0.42 X_C + 0.63. \quad (7.17)$$

Thus, the following Nu correlation is derived for a packed bed of shredded tires undergoing pyrolysis and gasification:

$$\text{Nu} = a_1(X_C) + a_2(X_C) \text{Re}^{a_3(X_C)} \text{Pr}^{a_4(X_C)}, \quad (7.18)$$

with a_1 to a_4 given by eqs. (7.14)–(7.17).

Nu correlations showed a decrease in Nu correlation during the pyrolysis and subsequent increase during gasification. These changes were associated

7.4. Heat transfer characterization

with morphological changes during the reaction: the evolution of highly porous particles during pyrolysis and the subsequent break-up of the particles. Former allows for more direct paths of the fluid through the packed bed and, therefore, reducing the accessible surface area for convective heat exchange between the two phases. In the initial and final state the packed bed is composed of rigid particles forcing the fluids around general increasing the accessible surface area for heat exchange.

For validation purposes several Nu correlations for packed beds of spherical and nonspherical particles, given in the literature [66, 225, 71, 187], are compared. Gnielinski's correlation [66] resulted from a combination of Nu correlations around a single sphere for laminar and turbulent flow,

$$\mathrm{Nu_{Gn}} = (1 + 1.5(1-\varepsilon))\mathrm{Nu_s},$$
$$\mathrm{Nu_s} = \mathrm{Nu_{s,min}} + \sqrt{\mathrm{Nu_{s,lam}}^2 + \mathrm{Nu_{s,turb}}^2},$$
$$= \left(2 + \sqrt{(0.664\mathrm{Re}^{0.5}\mathrm{Pr}^{1/3})^2 + \left(\frac{0.037\mathrm{Re}^{0.8}\mathrm{Pr}}{1 + 2.443\mathrm{Re}^{-0.1}(\mathrm{Pr}^{2/3}-1)}\right)^2}\right). \quad (7.19)$$

Wakao's correlation [225] was obtained by fitting experimental results, correcting for axial dispersion if needed,

$$\mathrm{Nu_{Wa}} = 2 + 1.1\mathrm{Re}^{0.6}\mathrm{Pr}^{1/3}. \quad (7.20)$$

Gunn's correlation [71] is based on a stochastic model for the packed bed geometry,

$$\mathrm{Nu_{Gu}} = (7 - 10\varepsilon + 5\varepsilon^2)(1 + 0.7\mathrm{Re}^{0.2}\mathrm{Pr}^{1/3})$$
$$+ (1.33 - 2.4\varepsilon + 1.2\varepsilon^2)\mathrm{Re}^{0.7}\mathrm{Pr}^{1/3}. \quad (7.21)$$

Saidi's correlation [187] is derived from fitting experimental results with a packed bed of non-spherical particles with $\varepsilon = 0.8$ to a continuum model:

$$\mathrm{Nu_{Sa}} = 0.015 + 0.11\mathrm{Re}^{0.75}\mathrm{Pr}^{0.75}. \quad (7.22)$$

The Nu correlation determined for the reacting packed bed of complex, porous, and non-spherical particles, given by eq. (7.18), is compared to those Nu correlations given by the models of eqs. (7.19)–(7.22). Results of this comparison are shown in figure 7.15 as a function of Re and for $\mathrm{Pr} = 0.1$ and 1. For the initial packed bed, the Nu correlations of eqs. (7.19)–(7.21) are not appropriate, especially for low Re. Mousa [143] reports measured Nu numbers exceeding 40 for $9 < \text{Re} < 30$ and $\text{Pr} = 0.7$. For the packed bed after pyrolysis (char), consisting of non-spherical and highly porous particles, the Nu numbers lie in the range proposed by Saidi [187] (RMS = 1.2 for Re·Pr < 25). For the final ash, all the modeled Nu correlations of eqs. (7.19)–(7.21) approach the one derived in this study. Ash particles are more spherical-like and less porous. The strong increase in Nu for the packed bed at $X_C = 0$ and 1 at high Re is related to the large Dupuit-Forchheimer coefficient and the larger tortuosities (see following sections), allowing for superior heat transfer.

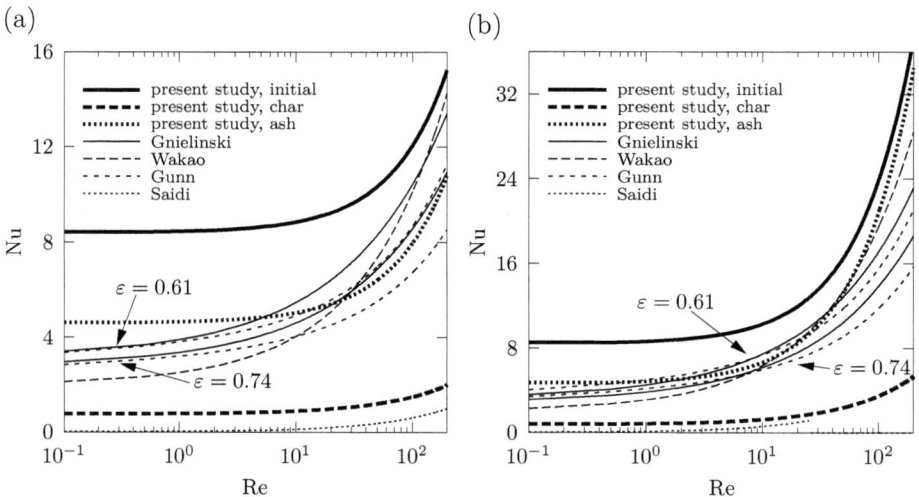

Figure 7.15: Nu correlations for packed beds given by Gnielinski ($\varepsilon = 0.61$ and 0.74) [66], Wakao [225], Gunn ($\varepsilon = 0.61$ and 0.74) [71], Saidi [187], and of the present study for $\text{Pr} = 0.1$ (a), and for $\text{Pr} = 1$ (b).

Packed beds of IOOS ($\varepsilon = \exp(-n\pi d^3/6)$) with corresponding particle sizes and porosities (see table 7.4) are generated to investigate on the influence of sphere size distribution and sphericity on the effective properties. DPLS is used for the determination of the effective properties. The numerically determined and fitted Nu correlations are depicted in table 7.7.

7.5 Mass transfer characterization

7.5.1 Permeability and Dupuit-Forchheimer coefficient

The sample size and the mesh and iteration convergence criteria used for the determination of the heat transfer coefficient are again used for the calculations

7.5. Mass transfer characterization

Table 7.7: Numerically determined Nu correlations for the three packed beds composed of IOOS with the same porosity and diameter as the reacting packed bed at $X_C = 0$, 0.68 and 1.

	Corresp. X_C	Nu	RMS
IOOS ($\varepsilon = 0.61$, $d=1.14$ mm)	0	$4.61 + 0.36Re^{0.699}Pr^{0.538}$	0.04
IOOS ($\varepsilon = 0.74$, $d=0.97$ mm)	0.68	$3.80 + 0.38Re^{0.662}Pr^{0.513}$	0.04
IOOS ($\varepsilon = 0.65$, $d=0.42$ mm)	1	$2.85 + 0.53Re^{0.614}Pr^{0.534}$	0.05

of the velocity and pressure field in the packed bed. The pressure variations associated with the artificially abruptly changing flow patterns at the sample's inlet/outlet (see chapter 3) represent less than 1% of the pressure drop across the sample and are therefore negligible.

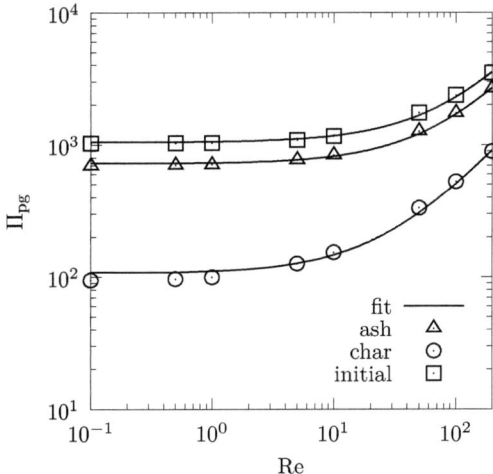

Figure 7.16: Dimensionless pressure gradient, Π_{pg}, as a function of Re for the three samples of the packed bed at $X_C = 0$, 0.68 and 1 (initial, char and ash).

The dimensionless pressure gradients for the three samples of the packed bed at different X_C are plotted in figure 7.16. The resulting K and F_{DP} are tabulated in table 7.8. The highest K of the packed bed is obtained at $X_C = 0.68$ (char). This is consistent with the fact that the highly porous, fractal-like particles evolved during pyrolysis. The lowest F_{DP} is obtained at $X_C = 0.68$. This again is explained by the porous structure of the particles contained in the bed, allowing the fluid to pass through the sample in a less disturbed manner,

Table 7.8: X_C-dependent permeability and Dupuit-Forchheimer coefficient of the reacting packed bed.

	K (m^2)	F_{DF} (m^{-1})	RMS
Initial ($X_C = 0$)	$1.242 \cdot 10^{-9}$	10941	43.4
Char ($X_C = 0.68$)	$8.817 \cdot 10^{-9}$	4130	14.9
Ash ($X_C = 1$)	$2.444 \cdot 10^{-10}$	23348	28.6

as is verified by the tortuosity distributions calculated in the next section. The following correlations for K,

$$K(X_C) = -4.03 \cdot 10^{-8} X_C^2 + 3.92 \cdot 10^{-8} X_C + 1.24 \cdot 10^{-9}, \qquad (7.23)$$

and F_{DF},

$$F_{DF}(X_C) = 75862 X_C^2 - 63255 X_C + 10941, \qquad (7.24)$$

of a packed bed of shredded tire particles undergoing pyrolysis and gasification are proposed.

The calculated K increases and decreases with increasing X_C while F_{DF} decreases and reincresases with increasing X_C. These changes were associated with morphological changes during the reaction: the evolution of highly porous particles during pyrolysis and the subsequent break-up of the particles. Former allows for more direct paths of the fluid through the packed bed and, therefore, reducing the pressure drop and the inertial resistance leading to higher K and lower F_{DF}. In the initial and final state the packed bed has lower porosities and is composed of rigid particles forcing the fluids around general increasing the pressure drop as well as the inertial resistance.

For validation purposes the obtained permeability and Dupuit-Forchheimer coefficient are compared to K and F_{DF} of packed beds of spherical and non-spherical particles available in the literature. K of a medium composed of parallel channels with Hagen-Poiseuille flow is given by [102],

$$K_{cap} = \frac{\varepsilon d^2}{32}. \qquad (7.25)$$

The hydraulic radius model [102] based on the Carman-Kozeny theory is given by

$$K_{hyd} = \frac{\varepsilon^3 d^2}{36 k_0 \tau (1-\varepsilon)^2} = \frac{\varepsilon^3 d^2}{36 k_K (1-\varepsilon)^2}, \qquad (7.26)$$

whereby k_K is the Kozeny constant, which is the product of a shape parameter, k_0, and the tortuosity, τ. k_K is approximated to be 5 for packed beds [102]. For

7.5. Mass transfer characterization

fibrous beds, Kyan et al. [113] proposed,

$$k_{K,\text{fib}} = \frac{\left(62.3\left(\sqrt{\frac{2\pi}{1-\varepsilon}} - 2.5\right)^2 (1-\varepsilon) + 107.4\right)\varepsilon^3}{16\varepsilon^3(1-\varepsilon^4)}. \tag{7.27}$$

Kuwabara, Sparrow et al., and Happel et al. [112, 198, 76] have derived correlations for k_K by solving the external flow around parallel and perpendicular arranged cylinders and spheres,

$$k_{K,\parallel} = \frac{2\varepsilon^3}{(1-\varepsilon)\left(2\ln\left(\frac{1}{1-\varepsilon}\right) - 3 + 4(1-\varepsilon) - (1-\varepsilon)^2\right)}, \tag{7.28}$$

$$k_{K,\perp} = \frac{\frac{2\varepsilon^3}{1-\varepsilon}}{\frac{1}{1-\varepsilon} - \frac{1-(1-\varepsilon)^2}{1+(1-\varepsilon)^2}}. \tag{7.29}$$

Rumpf et al. [185] showed that for a packed bed of spherical particles with a narrow size distribution, K is well approximated by

$$K_R = \frac{\varepsilon^{5.5}d^2}{5.6}. \tag{7.30}$$

Itoh [95] determined the permeability of a random array of rigid spheres accounting for the so called intermediate layer, a particular state formed around a test sphere. Davis et al. [44] calculated the permeability of a packed bed made of porous particles with an inner permeability, K_{inner}.

For higher Re numbers, MacDonald et al. [126] suggested a formulation to determine F_{DP} based on Ergun's theory,

$$F_{\text{DF,MD}} = 1.8\frac{1-\varepsilon}{\varepsilon^3 d}. \tag{7.31}$$

Ward [226] proposed

$$F_{\text{DF,W}} = \frac{0.55}{\sqrt{K}}. \tag{7.32}$$

Figure 7.17 compares the DPLS-determined K and F_{DP} of the reacting packed bed at $X_C = 0$, 0.68, and 1 (initial, char, and ash) with those obtained by eqs. (7.25)–(7.32) and models given by Itoh [95] and Davis [44]. For clarity, a packed bed with $K_{\text{inner}} = 10^{-10}$ m^2 is assumed for the calculations. ε and d (particle's diameter for all models) are adapted according to the sample specification for $X_C = 0$, 0.68, and 1. The values for K and F_{DP} scatter in a range of several orders of magnitude, spanned by the models.

Permeability and Dupuit-Forchheimer of the corresponding IOOS are depicted in table 7.9. The porosity affects significantly K and F_{DP}. This is qualitative in agreement with the results of the complex reacting packed bed.

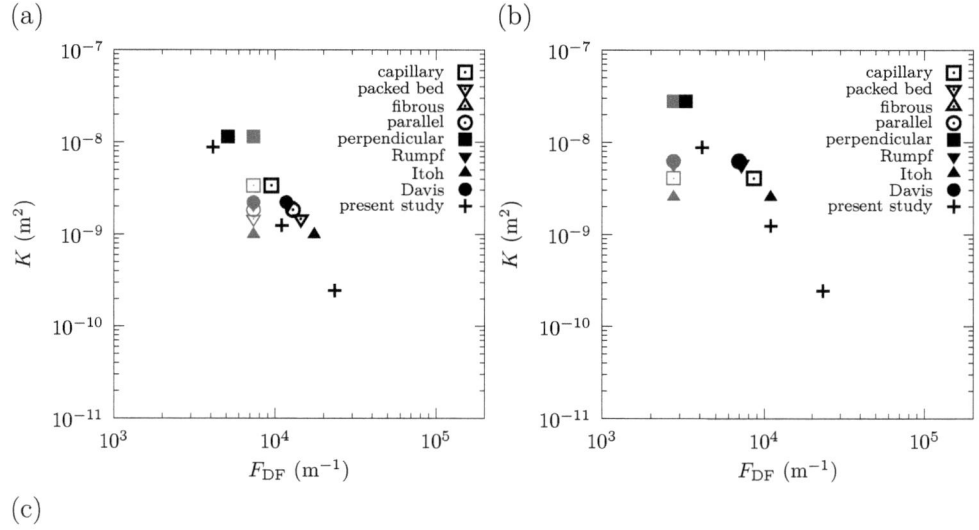

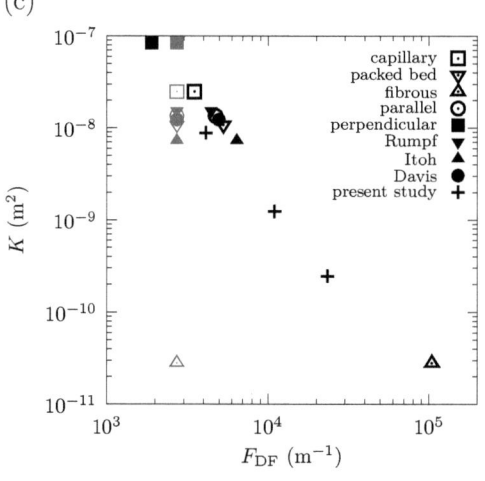

Figure 7.17: K and F_{DP} numerically determined by DPLS with those calculated by models for packed beds with $K_{inner} = 10^{-10}$ m^2 and for: $\varepsilon = 0.61$, $d = 1.14$ mm (a); $\varepsilon = 0.74$, $d = 0.97$ mm (b); $\varepsilon = 0.65$, $d = 0.42$ mm (c). F_{DP} is calculated by MacDonald (gray symbols, eq.(7.31)) and by Ward (black symbols, eq.(7.32)).

7.5. Mass transfer characterization

Table 7.9: Permeability and Dupuit-Forchheimer coefficient of the three packed beds composed of IOOS with the same porosity and diameter as the reacting packed bed at $X_C = 0$, 0.68 and 1.

	Corresp. X_C	K (m^2)	F_{DP} (m^{-1})	RMS
IOOS ($\varepsilon = 0.61$, $d=1.14$ mm)	0	$1.581 \cdot 10^{-8}$	2041	5.23
IOOS ($\varepsilon = 0.74$, $d=0.97$ mm)	0.68	$3.352 \cdot 10^{-8}$	779	1.88
IOOS ($\varepsilon = 0.65$, $d=0.42$ mm)	1	$3.451 \cdot 10^{-9}$	3104	4.05

Table 7.10: Mean, mode and median tortuosity and residence time for the packed bed at Re = 1.

Sample	τ_m	τ_{mode}	τ_{median}	t_m (s)	t_{mode} (s)	t_{median} (s)
$X_C = 0$	1.256	1.184	1.232	0.014	0.011	0.010
$X_C = 0.68$	1.196	1.094	1.165	0.030	0.006	0.009
$X_C = 1$	1.204	1.150	1.197	0.003	0.002	0.003

7.5.2 Tortuosity and residence time

Calculated distributions of tortuosity, τ, and residence time, t, for the packed bed at $X_C = 0$, 0.68, and 1 (initial, char, and ash) and Re = 1 are shown in figure 7.18. Mean, median, and mode of the calculated distributions of τ and t at Re = 1 are given in table 7.10. The following correlations of τ_m and t_m for a packed bed of shredded tire particles undergoing pyrolysis and gasification are derived (Re = 1, $l_{sample} = 2.3$ mm):

$$\tau_m(X_C) = 0.08X_C^2 - 0.14X_C + 1.26, \tag{7.33}$$
$$t_m(X_C) = -0.12X_C^2 + 0.10X_C + 0.01. \tag{7.34}$$

τ at $X_C = 0.68$ is smaller than the one at $X_C = 0$ and 1. Since the particles after pyrolysis are highly porous, fluid is able to pass through the packed bed with lower tortuosity. The τ and t distributions at $X_C = 0.71$ exhibit a tail accounting for fluid particles trapped for some time at unconnected pores, resulting in larger residence times.

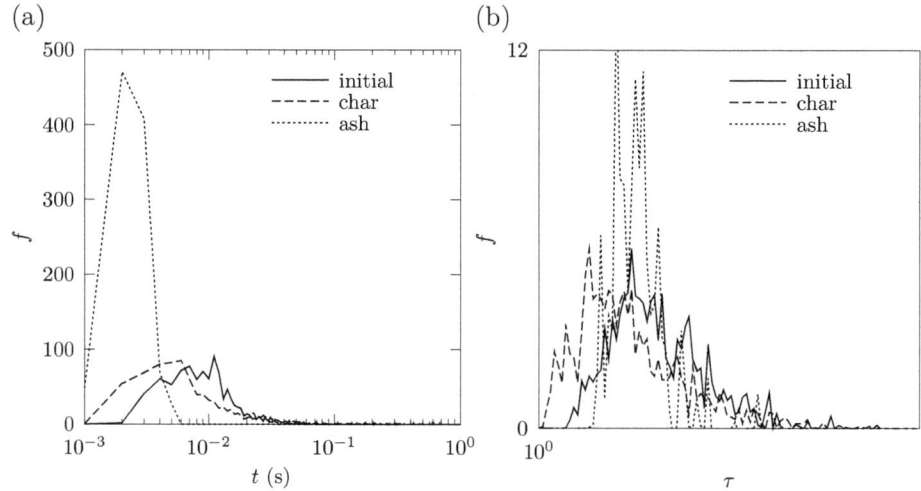

Figure 7.18: Residence time (a) and tortuosity (b) distribution at Re = 1 for the reacting packed bed at $X_C = 0$, 0.68, and 1 (initial, char, and ash).

7.6 Conclusions

Discrete-scale numerical simulation was applied to a reacting packed bed whose complex 3D geometry had been determined by CT at different reaction extents. The pyrolysis and gasification of carbonaceous material (shredded waste tire) to syngas was chosen as the model reaction.

The variation in the morphology during the thermochemical solid-gas reaction of the packed bed was experimentally investigated at discrete carbon conversion steps ($X_C = 0$, 0.68, 0.79, 0.9, and 1) and for different process parameters (feedstock size, furnace temperature, gasifying agent, and partial pressure of gasifying agent). Porosity (measured by weight measurements), BET surface area (measured by N_2 adsorption), and particle-size distribution (measured laser scattering) were compared to numerically determined morphological characteristics. Discrepancies were explained by limitations in the CT scan resolutions and to image distortions around optically thick heavy metal impurities. The morphological results can be used for the determination of structural parameters needed in kinetic models [125].

The effective radiative heat transfer properties (extinction coefficients, scattering albedo and scattering phase function) were determined as a function of reaction extent. The extinction coefficient increased as particles shrank and

7.6. Conclusions

shortened the attenuation path length. For diffusely reflecting particles, the scattering phase function was found to be independent of the reaction extent. For specularly reflecting particles, the scattering phase function exhibited a strong forward peak and dependency on the refractive index, and therefore X_C.

The effective conductivity of the packed bed was calculated, neglecting contact resitance at the particle-particle boundary. It decreased as the reaction progresses due to increased porosity (during pyrolysis) and due to particle shrinkage (during gasification). k_e/k_s converges to 0.0114, 0.0056, and 0.002 for $X_C = 0$, 0.68 and 1, respectively, for low k_f/k_s values.

The convective heat transfer coefficient, permeability, and Dupuit-Forchheimer coefficient of the reacting packed bed were determined and compared to correlations available in the literature. The initial packed bed was characterized by larger Nu numbers than those obtained with the models of Gnielinski or Wakao [66, 225] but smaller values than those measured by Mousa [143]. The packed bed after pyrolysis (char) was composed of porous and nonspherical particles. The final packed bed (ash) showed Nu values comparable to those proposed for packed beds of rigid and spherical particles. The packed bed after pyrolysis (char) showed the highest K and the lowest F_{DF}. This was because the high inner porosity of particles enhanced the bed porosity and shortened the calculated tortuosity.

All effective transport properties were found to be strong functions of the reaction extent. Analytical β, β/σ_s, Φ_d, Φ_{sp}, k_e/k_s, Nu, K, F_{DF}, τ_m and t_m correlations for a packed bed undergoing pyrolysis and gasification were derived as a function of the reaction extent.

These effective transport properties can in turn be incorporated in volume-averaged (continuum) models for the purpose of design and optimization of (solar) packed-bed reactors used for gasification processes.

Chapter 8
Semitransparent-particle packed bed[1]

In this chapter, the morphological and spectral radiative characteristics of a packed bed composed of nonspherical, semitransparent particles are determined. As the radiative properties of complex morphological structures significantly deviate from those obtained when applying simplifying approximations, e.g., spherical particles, CT is applied in order to obtain the exact 3D geometrical representation of the complex porous media. A packed bed of $CaCO_3$ particles, used in industrially relevant high-temperature processes and in capturing CO_2 from combustion flue gases or air [201, 148], is used as model reaction and introduced in the first part of this chapter.

Thermal decomposition of $CaCO_3$ takes place at reasonable rates above 1000 K. At these temperatures, thermal radiative transport becomes the dominant heat transfer mode. Thus, knowledge of the radiative properties of packed beds is crucial for the engineering design and optimization of these chemical reactors and processes. The CT-based radiative characteristics of the packed bed of semitransparent particles are determined in the second part of this chapter.

Experimental investigations of transmittance and extinction coefficients have been performed for packed beds of opaque, transparent, and semitransparent phases [30, 98, 156]. Numerical simulation were carried out for artificially generated packed beds of spherical particles [213] by applying the method of MC-based RDFs [211], MC [36, 37], two-flux [236, 26, 8], and DO [24] methods. In contrast to previous investigations, the CT-based methodology enables the determination of accurate, effective radiative properties in the limit of geometric optics with negligible diffraction, which, in turn, can serve as reference values to those obtained by approximate numerical or experimental methods.

---

[1]Material from this chapter has been published in: S. Haussener, W. Lipiński, J. Petrasch, P. Wyss, and A. Steinfeld. Tomographic characterization of a semitransparent-particle packed bed and determination of its thermal radiative properties. *Journal of Heat Transfer*, 131: 072701, 2009. [82]

8.1 Thermal decomposition of calcium carbonate

The high-temperature processing of calcium carbonate (limestone) leads to the production of calcium oxide widely used in construction and building, steel making processes, agriculture and food industry, water treatment, and fine chemical production. It also describes the high-temperature step of a thermochemical multi-step process proposed for CO_2 capturing from flue gases and air [201, 148]. It is described by the simplified net reaction:

$$CaCO_3 = CaO + CO_2. \qquad (8.1)$$

The reaction proceeds endothermically above 1200 K at 1 atm. The reaction enthalpy is 270 kJmol^{-1} when the reactants are initially at 298 K and the products at 1200 K.

Solar energy instead of fossil fuels has been proposed as energy to drive reaction (8.1), which results in a CO_2 reduction potential of 20 to 40%. Economical evaluations have predicted lime prices of 128 to 157 \$/t for the solar production of lime in 2005 [136]. This is about two or three times the selling price of conventionally produced lime.

A solar reactor, based on an indirectly irradiated multi-tube reactor [80], has been proposed [135]. The reactor is depicted in figure 8.1. Solar radiation enters the reactor and heats the cavity and the absorber tubes, which transport the energy to the small grained calcium carbonate particles. The particles continuously enter the rear of the reactor and are pre-heated before being transported through the absorber tube, where they react.

As the reactor is rotating moderately, the particles build packed beds within the absorber tubes. Packed beds provide relatively high contacting area allowing for efficient heat transfer. The radiative characterization of the semitransparent particle bed is of importance for design and optimization of the reactor and process.

8.2 Morphological characterization

8.2.1 Computed tomography

The sample investigated consists of a packed bed of nonspherical $CaCO_3$ particles (source: Carrara marble, Ferret's diameter $\approx$ 3 mm) randomly placed in a 4.5 cm-diameter rubber tube, as seen in figure 8.2.a. The sample is exposed to a polychromatic X-ray beam, generated by electrons incident on a wolfram target and filtered by a 0.0025 mm Re filter. The generator is operated at an accel-

8.2. Morphological characterization

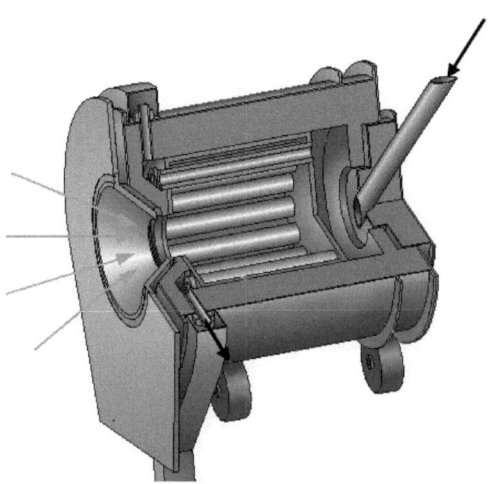

Figure 8.1: Schematic of the indirectly irradiated solar calcination reactor [135]. The solar radiation is absorbed by the tubes, which emit the radiation to the calcium carbonate particles. The particles are transported through the tube and react. The products leave the reactor at the front side.

eration voltage of 140 keV and a current of 0.11 mA. A Hamamatsu flatpanel C7942 CA-02 protected by a 0.1 mm brass filter is used to detect the transmitted X-rays. The sample is scanned at 600 angles (projections). Each projection is an average of 8 scans with an exposure time of 0.25 s, leading to a voxel size of 45 µm. Figures 8.2.b and 8.2.c show the 2D tomographic image and the 3D surface rendering of the sample, respectively. The histogram of the normalized absorption values α/α_{max}, shown in figure 8.3, reveals two distinct peaks that account for the solid and fluid (void) phases. The calculated and nearly identical width of the two peaks supports the assumption of homogeneity. The calculated minimum between the two peaks ($\alpha/\alpha_{max} = 0.43$) is used as threshold value for phase identification. Phase boundaries between two $CaCO_3$ particles are neglected, as they cannot be distinguished in the tomography images.

8.2.2 Porosity and specific surface

Figure 8.4.a shows the two-point correlation function for the $CaCO_3$ packed bed. The calculated porosity is 0.39, which compares well with the porosity determined experimentally from the sample weight 0.40 ± 0.02. This value is close to the one for a dense random or orthorhombic packed bed made of uniform

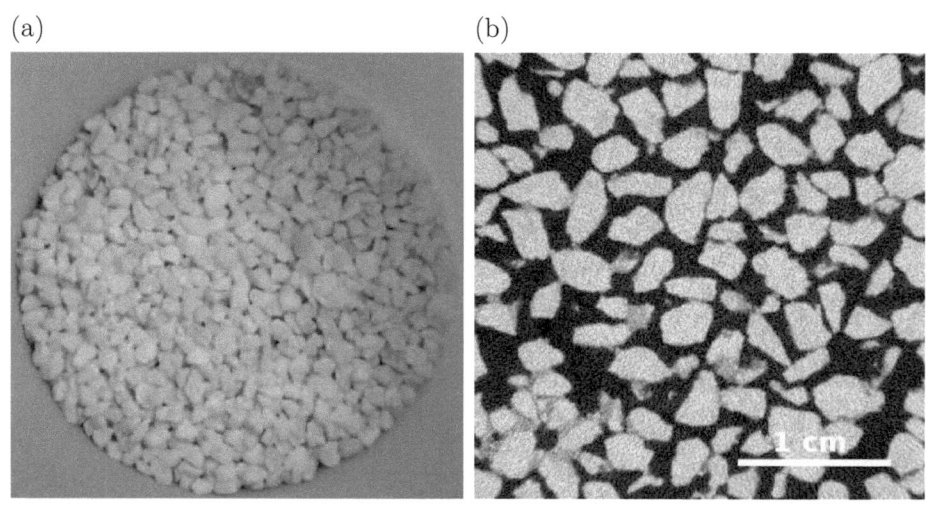

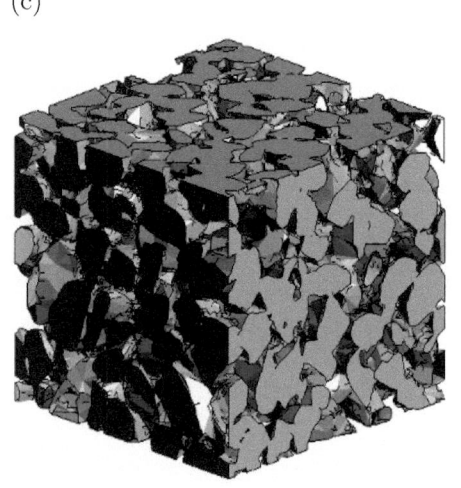

Figure 8.2: Sample of the packed bed of $CaCO_3$ particles: top view photograph (a), 2D tomographic image (b), and 3D surface rendering (c).

8.2. Morphological characterization

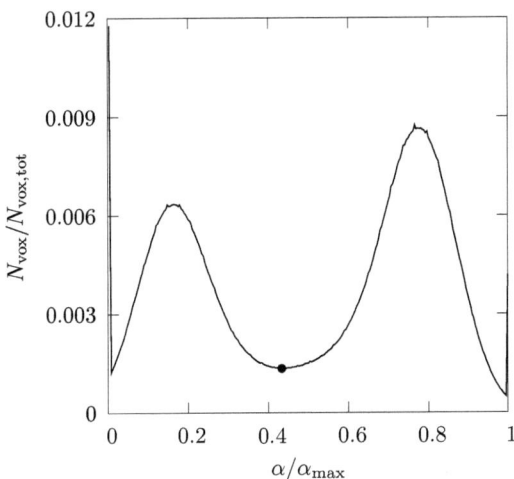

Figure 8.3: Normalized histogram of the sample's absorption values obtained by CT for the void phase (left peak) and for the solid phase (right peak). The bullet indicates the threshold value $\alpha/\alpha_{\max} = 0.43$ used for phase identification.

monodispersed spheres [235]. The calculated specific surface area is $1.31 \cdot 10^3$ m^{-1}, which corresponds to an analytically calculated particle diameter of 2.8 mm for uniform monodispersed spheres. As will be shown later, the hydraulic particle diameter for the randomly shaped $CaCO_3$ particles is approximately 1.9 mm.

8.2.3 Representative elementary volume

REV is determined based on the porosity for subvolume sizes at ten random locations in the sample, as indicated in figure 8.4.b. Note that for an edge length l approaching 0, the porosity is either 0 or 1, depending on whether the point lies in the void or in the solid phase. Assuming a tolerance band of ± 0.01, l_{REV} = 0.0152 m (for ± 0.02, l_{REV} = 0.0105 m; for ± 0.05 l_{REV} = 0.0054 m) and is used as the minimum sample size in the following analysis.

8.2.4 Pore- and particle-size distributions

An opening with a spherical structuring element, is applied to calculate the pore- and particle-size distributions, i.e., the size distribution defined by the sphere that fits completely within the pore or particle space, respectively. Figure 8.5 shows the distribution functions of the pore and particle sizes. The mean, mode,

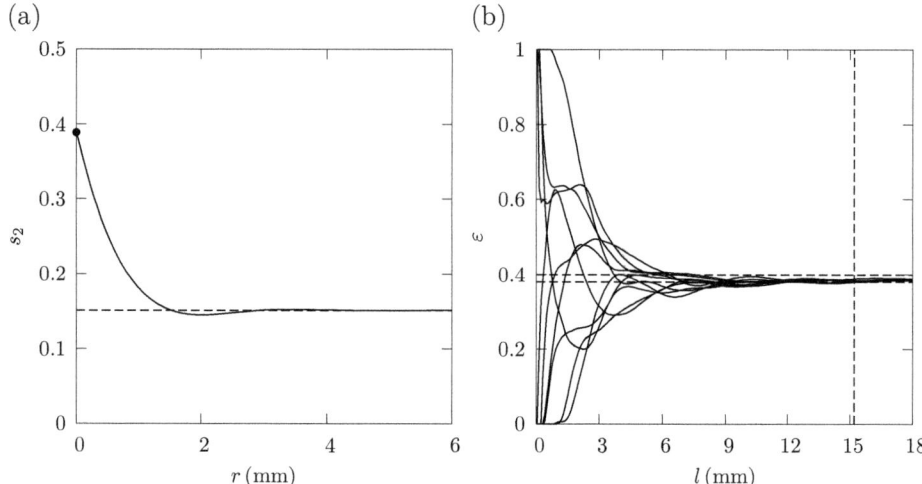

Figure 8.4: (a) Two-point correlation function for the CaCO$_3$ packed bed. The value at $r = 0$ corresponds to the bed porosity. The dashed line indicates the asymptotic value of the function, which corresponds to ε^2. (b) Determination of the REV edge length (indicated by the vertical dashed line) by calculating the porosity of ten subvolumes with varying edge lengths l at random locations. The tolerance band for conversion and determination of the REV volume at $\varepsilon \pm 0.01$ is indicated by the two horizontal dashed lines.

median, and hydraulic diameters are listed in table 8.1. Thus, the assumption of geometrical optics is valid for $\lambda < 3500$ µm.

8.3 Radiative Properties

In this section, the void phase and the CaCO$_3$ particles of the packed bed are referred to as fluid and solid phases, respectively. The corresponding phase indices i, j used in the governing equations and methodology (see chapters 2 and 3) are 1 for the fluid phase and 2 for the solid phase. Spectral calculations of the radiative properties are performed for 150 distinct points between 0.1 and 100 µm. Reflection and refraction at the specularly reflecting and diffusely reflecting interfaces are modeled by Fresnel's equations and diffuse reflection/refraction, respectively.

8.3. Radiative Properties

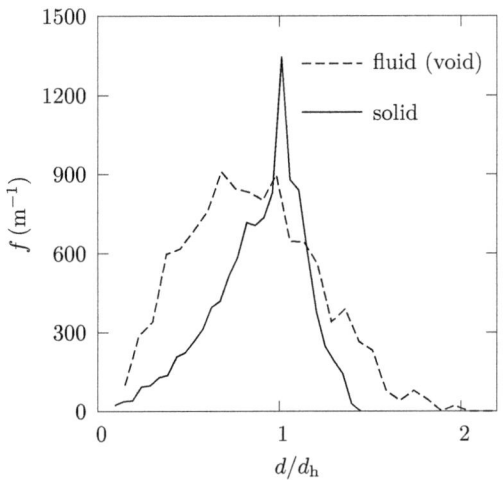

Figure 8.5: Opening size distribution functions of the solid and fluid phases of the $CaCO_3$ packed bed ($d_h = d_{h,pore}$ for fluid and $d_h = d_{h,particle}$ for solid).

Table 8.1: Arithmetic mean diameter, mode, median, and hydraulic diameter calculated from the pore and particle size distributions.

	Pore	Solid
d_m (mm)	1.00	1.66
d_{mode} (mm)	0.81	1.89
d_{median} (mm)	0.98	1.75
d_h (mm)	1.19	1.87

8.3.1 Single-phase internal radiative properties

The fluid phase is assumed to be transparent, i.e., its internal absorption and scattering coefficients $\sigma_{s,1}$ and κ_1, respectively, are equal to 0, and its refractive index is equal to 1. The bulk properties of $CaCO_3$ are determined based on the properties of $CaCO_3$ grains (monocrystals). They are randomly shaped and oriented, as seen in the scanning electron microscope (SEM) photograph shown in figure 8.6. Their characteristic size is 160 μm. Figure 8.7.a shows the complex refractive index $m_2 = n_2 - ik_2$ of $CaCO_3$ [158]. In the spectral range 0.2 to 6 μm, the imaginary part was obtained by applying the Lorentz theory [174]. The directional-hemispherical reflectivities at the specular fluid-solid interface for radiation incident from the fluid phase $\rho_{r,sp,12}$ and for radiation in-

Figure 8.6: SEM picture of a single $CaCO_3$ particle.

cident from the solid phase $\rho_{r,sp,21}$ are calculated using Fresnel's equations [141]. The directional-hemispherical reflectivity at the phase boundary as a function of the incident angle is exemplary shown in figure 8.8.a for changing n_1/n_2 at constant $k_i = 0$. $\rho_{r,sp,ij}$ stays nearly constant at small incident angles, where its value increases with increasing n_1 for $n_1 > n_2$ or increasing n_2 for $n_2 > n_1$, respectively. A steep increase of $\rho_{r,sp,ij}$ at large incident angles ($\theta_{in} > 70$ °) is observed for $n_2 > n_1$ and an abrupt increase at the total reflection angle for $n_1 > n_2$. Changes in k_i (for $k_i < 1$) do not have a significant influence on the directional-hemispherical reflectivity. The refraction angle at the phase boundary, calculated by Fresnel's equations [141], as a function of the incident angle is given in figure 8.8.b. The maximum refraction angle from medium 1 to medium 2, with $n_1 < n_2$, is decreasing with increasing n_2 at constant $k_i = 0$. The total reflection angle from medium 1 to medium 2, with $n_1 > n_2$, is decreasing with increasing n_1 at constant k_i. The imaginary part of the refractive index has a minor influence on the refraction angle for small k_i values. The hemispherical reflectivities of both sides of a diffuse fluid-solid interface $\rho_{r,d}$ are equal to that for a diffuse $CaCO_3$ surface [180]. For $\lambda < 0.8$ µm and $\lambda > 2.5$ µm, the hemispherical reflectivity is extrapolated with constant values 0.85 and 0.87, respectively. $\rho_{r,sp,12}$, $\rho_{r,sp,21}$ and $\rho_{r,d}$ are shown in figure 8.7.b as a function of wavelength. The internal radiative coefficients of $CaCO_3$, $\sigma_{d,s,1}$, and $\kappa_{d,2}$, obtained by applying the Mie theory, are shown in figure 8.9.a [16]. The spectral oscillations, particularly for $\lambda > 6$ µm, result from oscillations in the complex re-

8.3. Radiative Properties

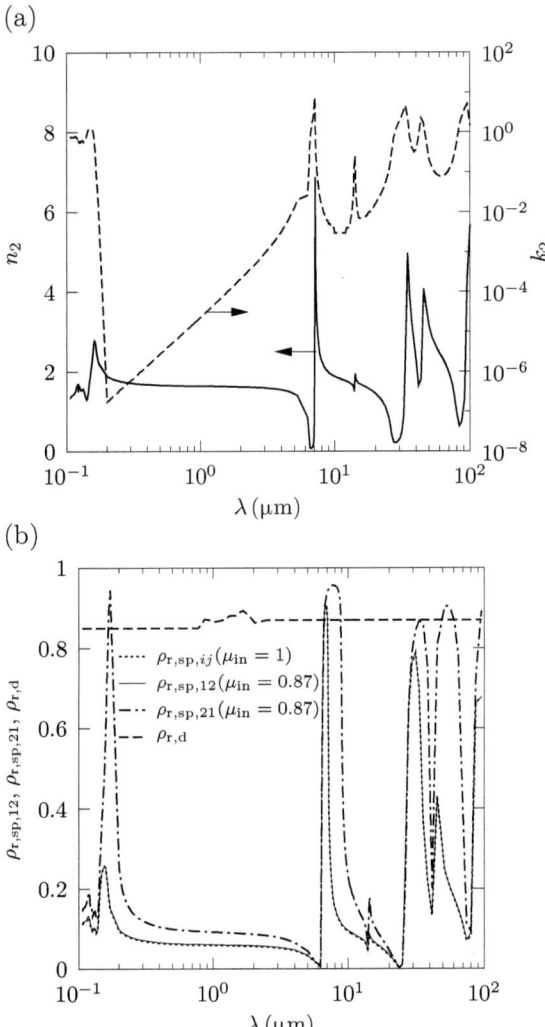

Figure 8.7: (a) Complex refractive index of $CaCO_3$: real part obtained experimentally (solid line) [158], and imaginary part obtained by the Lorentz theory in the spectral range 0.2 to 6 µm [174] and experimentally in the remaining range (dashed line) [158]. (b) Spectral directional-hemispherical reflectivities at the specular fluid-solid interface for selected incidence directions, and spectral hemispherical reflectivity of the diffuse fluid-solid interface.

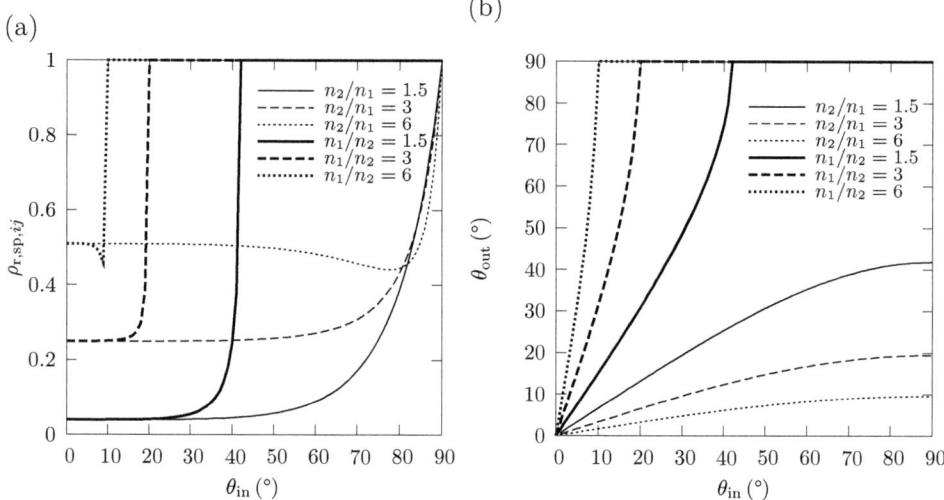

Figure 8.8: Directional-hemispherical reflectivity as a function of incident angle at phase boundary from medium 1 to medium 2: $k_i = 0$, with $n_1 < n_2$ or $n_1 > n_2$ (a), and refraction angle as a function of incident angle from medium 1 to medium 2: $k_i = 0$, with varying $n_1 < n_2$ or $n_1 > n_2$ (b).

fractive index (see figure 8.7). The internal scattering within $CaCO_3$ is assumed to be isotropic ($\Phi_{d,2} = 1$). Figure 8.9.b shows the ratio of the scattering efficiency factor for dependent scattering, calculated by gas, packed-sphere, liquid, or modified-liquid model [214], to that for independent scattering calculated by Mie theory (assumed $f_v = 1$). Dependent scattering effects are thus neglected in this study since the maximal reduction in the scattering efficiency factor for 100 µm $> \lambda >$ 0.1 µm is only approximately 10%.

8.3.2 Two-phase medium radiative coefficients

The scattering coefficients of the $CaCO_3$ packed-bed two-phase medium are shown in figure 8.10 as a function of wavelength for the fluid phase (figure 8.10.a and 8.10.b) and the solid phase (figure 8.10.c and 8.10.d), assuming specularly reflecting particles (figure 8.10.a and 8.10.c) and diffusely reflecting particles (figure 8.10.b and 8.10.d). The reflection behavior of the fluid-solid interface significantly influences $\sigma_{s,ij}$ and $\sigma_{s,\text{refl},i}$. For the fluid phase and specular interface, $\sigma_{s,\text{refl},i} < \sigma_{s,ij}$ for $\lambda < 6$ µm, while for the diffuse interface $\sigma_{s,\text{refl},i} > \sigma_{s,ij}$. For the solid phase, $\sigma_{s,\text{refl},i} > \sigma_{s,ij}$, and this trend is independent of the reflection

8.3. Radiative Properties

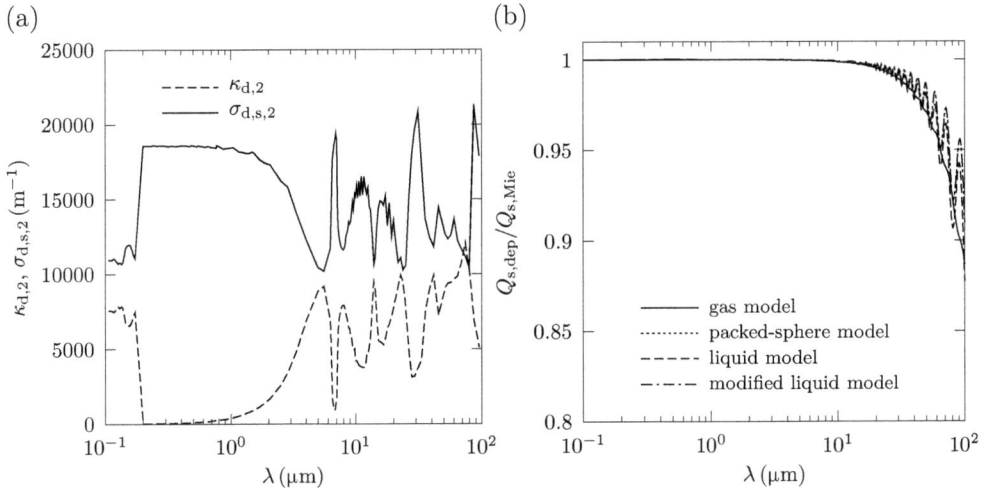

Figure 8.9: (a) Internal absorption and scattering coefficients of CaCO$_3$ particles. (b) Ratio of the scattering efficiency factor obtained for dependent scattering calculated by gas, packed-sphere, liquid, and modified-liquid models to the one obtained for independently scattering calculated by Mie theory.

interface type. $\sigma_{s,refl,1}$, obtained for the specular and diffuse interfaces, clearly follows the shape of $\rho_{r,sp}$ and $\rho_{r,d}$, respectively (see figure 8.7.b). $\sigma_{s,refl,2}$ and $\sigma_{s,21}$ are additionally influenced by the presence of the total reflection phenomenon in the particle. Obviously, for any phase, $\sigma_{s,ij}$ is complementary to $\sigma_{s,refl,i}$ since it refers to the transmitted portions of radiation across the interface.

The extinction coefficients β_i are shown in figures 8.11. They are independent of the interface reflection type (specular/diffuse). In addition, β_1 is independent of λ as it is a function of the interface geometry only. β_2 increases with λ because of the increasing $\kappa_{d,2} + \sigma_{d,s,2}$. The spectral oscillations in β_2 result from the statistical MC oscillations. The spectral oscillations in β_2 are mostly the result of the spectral oscillations of $\kappa_{d,2}$ and $\sigma_{d,s,2}$ (see figure 8.9.a).

8.3.3 Two-phase medium scattering phase functions

The probability density functions of the directional cosine of the incidence angle at the fluid-solid interface, determined by MC, are plotted in figure 8.12 for both phases. Assuming $\kappa_{d,2} = \sigma_{d,s,2} = 0$, $F_{\mu_{in},1}$ and $F_{\mu_{in},2}$ compare well to $F_{\mu_{in}}$, computed for identical overlapping opaque spheres (IOOS) and for identical overlapping transparent spheres (IOTS), respectively [211]. For the real values

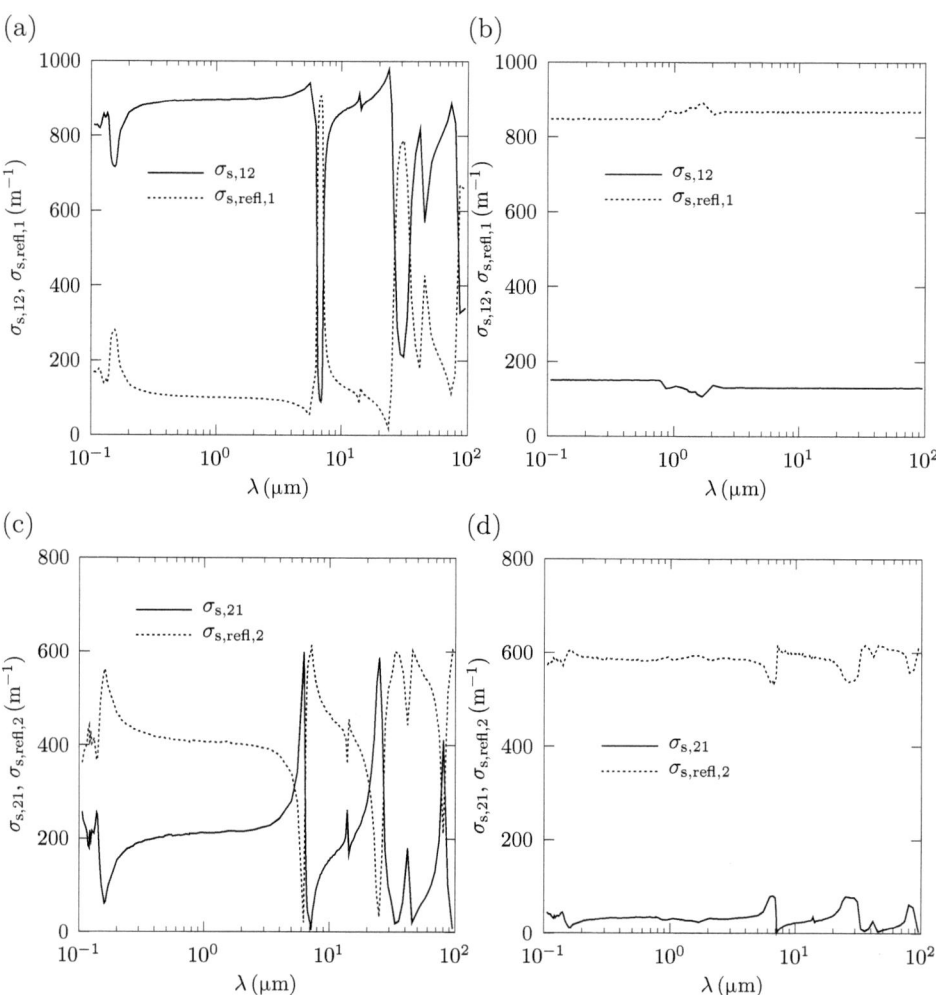

Figure 8.10: Spectral scattering coefficients of the CaCO$_3$ packed bed for the fluid phase (a,b) and the solid phase (c,d), assuming specularly reflecting particles (a,c) and diffusely reflecting particles (b,d).

8.3. Radiative Properties

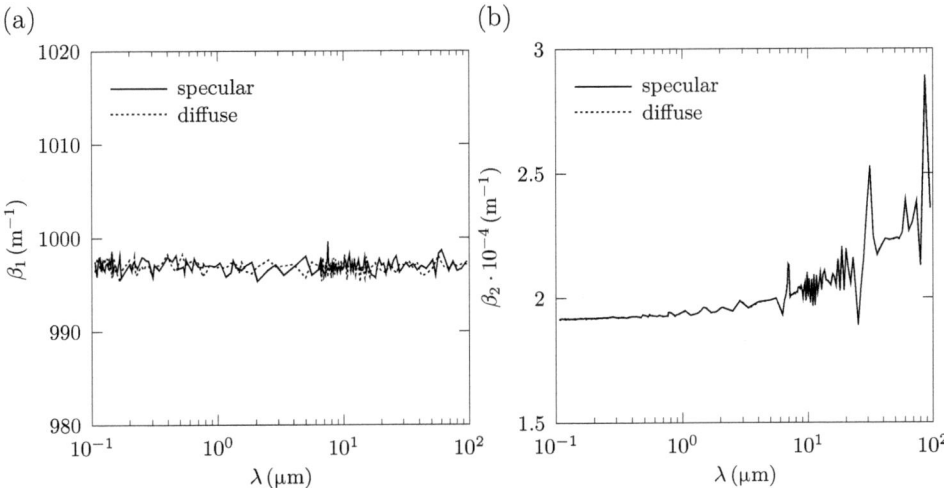

Figure 8.11: Spectral extinction coefficients of the packed bed: fluid phase (a) and solid phase (b).

of $\kappa_{d,2}$ and $\sigma_{d,s,2}$, $F_{\mu_{in},2}$ depends weakly on β_2 and hence on λ. The dependency of $F_{\mu_{in},2}$ on β_2 is explained by the fact that, for increasing β_2, the incidence angles corresponding to longer paths occur less frequently.

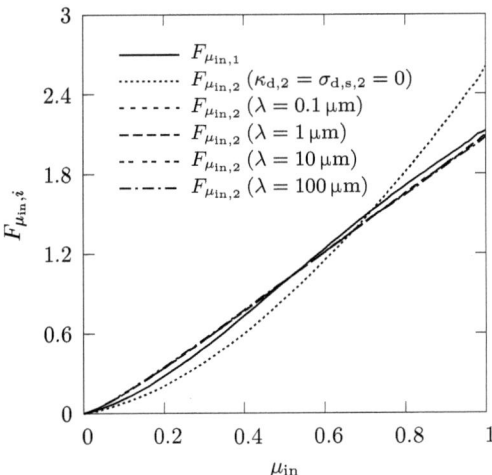

Figure 8.12: Probability density functions of the directional cosine of the incident angle at the fluid-solid interface for selected wavelengths.

Functions $F_{\mu_{\text{in}},i}$ are then used to calculate the scattering phase functions, see chapter 3. Figures 8.14 and 8.13 show the scattering phase functions for spectral and diffuse interfaces, respectively, as a function of the cosine of the scattering angle at selected wavelengths $\lambda = 0.1$ µm, 1 µm, 10 µm, and 100 µm. Φ_{12} and Φ_{21} are identical for the specularly reflecting solid-fluid interface because the Fresnel's equation yields the same results when the angle of incidence and transmission are interchanged [141]. The forward scattering peak decreases with λ because of an increase in normal-hemispherical reflectivity of the interface, which in turn is the result of the increasing n_2 and k_2. Since total reflection is assumed for the diffuse interface, $\Phi_{21} = 0$ for $\mu_s < -\sqrt{1 - \mu_{\text{refl,tot}}^2}$. For the specular interface, $\Phi_{\text{refl},1}$ and $\Phi_{\text{refl},2}$ are clearly functions of λ due to the spectral variation of $\rho_{\text{r,sp}}$, while this cannot be observed for the diffuse interface. The jump of $\Phi_{\text{refl},2}$ is explained by enhanced interface reflections for $\mu_s > 1 - 2\mu_{\text{refl,tot}}^2$. Obviously, Φ_{12} is equal to $\Phi_{\text{refl},1}$ because of the absence of internal scattering in the fluid phase. In contrast, Φ_{22} is mostly influenced by $\Phi_{\text{d},2}$ because of the large internal scattering in the solid phase (see eq. (2.36)).

8.3.4 Sensitivity analysis

A MC parametric study is carried out to elucidate the influence of the uncertainties in n_2, k_2, and $\rho_{\text{r,d}}$ on the radiative properties by varying by $\pm 20\%$ their reference values $n_2 = 1.64$, $k_2 = 2.3 \cdot 10^{-5}$, and $\rho_{\text{r,d}} = 0.87$ at $\lambda = 1$ u m. These variations do not affect β_i (<1% change). κ_2 is influenced by variations of k_2 (up to 20%) and n_2 (6%) for specular and diffuse fluid-solid interfaces.

For the specular interface, variation of n_2 leads to remarkable effect on $\sigma_{\text{s,refl},1}$, $\sigma_{\text{s,refl},2}$, $\sigma_{\text{s},12}$, and $\sigma_{\text{s},21}$ (up to 66% change), because $\rho_{\text{r,sp}}$ depends on n_2. For diffusely reflecting particles, variation of n_2 and k_2 leads to small variations in $\sigma_{\text{s,refl},1}$, $\sigma_{\text{s,refl},2}$, and $\sigma_{\text{s},12}$ only (up to 5% change) because of the varying particle optical thicknesses and the total reflection angles. The change in $\sigma_{\text{s},21}$ is up to 55%. The variation of $\rho_{\text{r,d}}$ has a more pronounced effect on the scattering coefficients. For example, for $\rho_{\text{r,d}} = 1.0$ (no refraction), $\sigma_{\text{s},ij} = 0$, while reduction in $\rho_{\text{r,d}}$ by 20% nearly doubles $\sigma_{\text{s},ij}$. Decreasing values of n_2 augments the forward scattering peaks of Φ_{12} and Φ_{21} for both specular and diffuse interfaces, as seen in figure 8.15.a. This is due to smaller relative angles between the incident and refracted rays for both specular and diffuse interfaces, according to Fresnel's equations.

Variations of k_2 and $\rho_{\text{r,d}}$ by $\pm 20\%$ do not affect the scattering phase functions. Variation of n_2 has an effect on $\Phi_{\text{refl},1}$ and $\Phi_{\text{refl},2}$ for the specular interface because the shape of $\rho_{\text{r,sp}}(\mu_{\text{in}})$ depends on n_2. Specular reflection generally leads to more

8.3. Radiative Properties

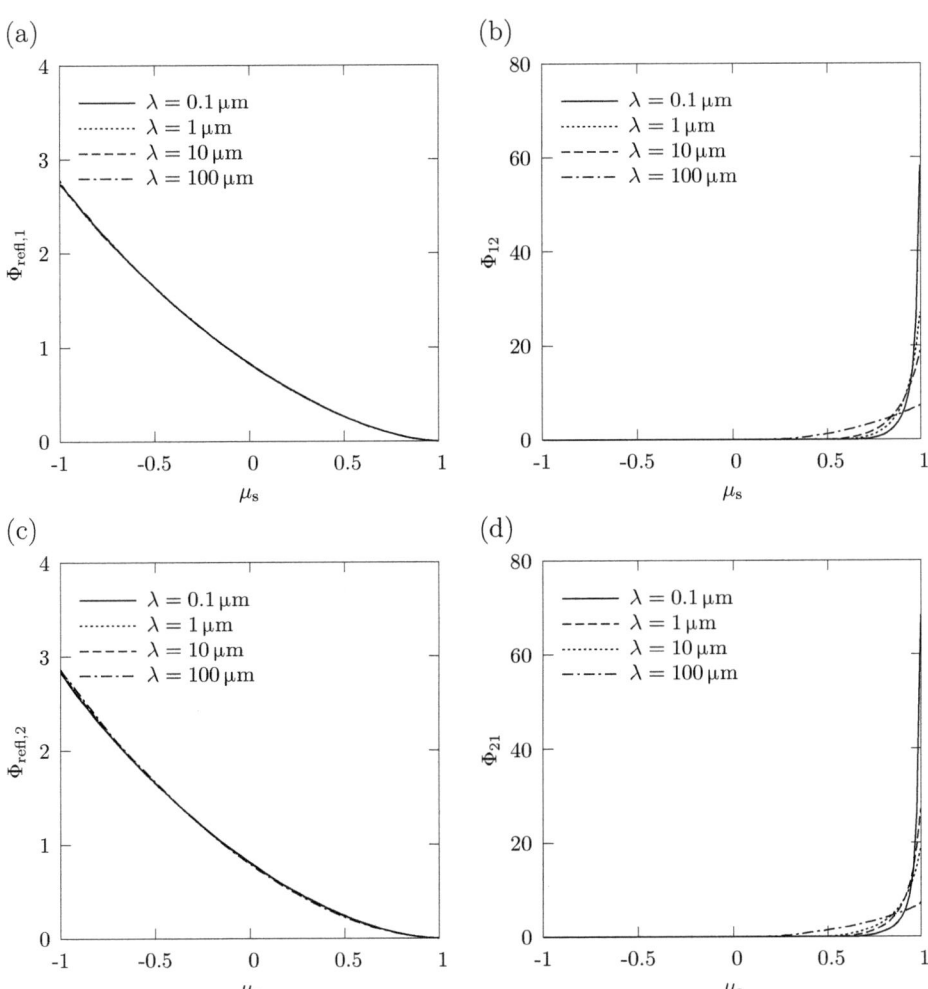

Figure 8.13: Scattering phase functions of the $CaCO_3$ packed bed versus cosine of the scattering angle for a diffusely reflecting solid-fluid interface, at selected wavelengths $\lambda = 0.1$ µm, 1 µm, 10 µm, and 100 µm.

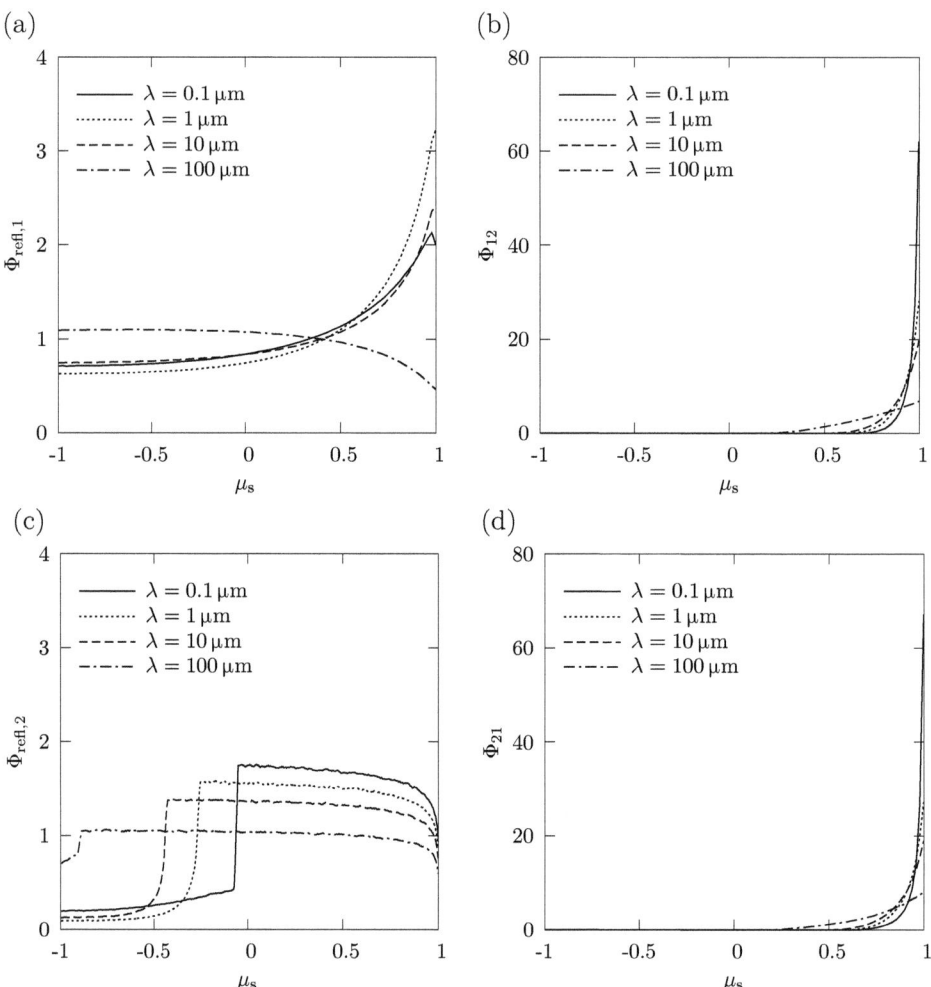

Figure 8.14: Scattering phase functions of the CaCO$_3$ packed bed versus cosine of the scattering angle for a specularly reflecting solid-fluid interface, at selected wavelengths λ = 0.1 µm, 1 µm, 10 µm, and 100 µm.

8.3. Radiative Properties

pronounced forward scattering in $\Phi_{\text{refl},1}$ and $\Phi_{\text{refl},2}$, mainly as a result of the enhanced reflective behavior of the interface at larger incidence angles, while backward scattering is favored for diffusely reflecting particles. This behavior is less evident for materials with higher β_2 and $\sigma_{\text{d,s},2}$, respectively, since the fraction of internal (isotropic) scattering is larger. Sensitivity analysis of the statistically determined characteristic size of the grains indicates that reduced grain size enhances the variations in the extinction behavior for wavelengths above 10 µm, while the opposite is true for larger grain sizes. Changing the characteristic size by ±20% leads to variations in the extinction efficiency of up to 13%, scattering efficiency of up to 25%, and absorption efficiency of up to 19%.

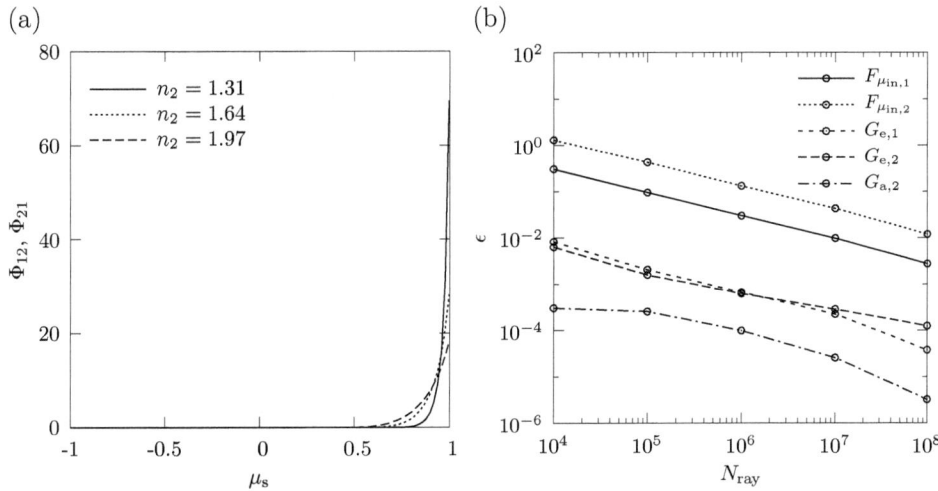

Figure 8.15: (a) Scattering phase function Φ_{12} and Φ_{21} for specular and diffuse fluid-solid interfaces, as a function of the scattering angle cosine, for selected refractive indices $n_2 = 1.31$, 1.64, and 1.97. (b) Normalized two-norm of the cumulative distribution functions.

8.3.5 Accuracy and validation of the MC algorithm

MC convergence is examined for 2-norms of $F_{\mu_{\text{in}},i}$, $G_{\text{e},i}$, $G_{\text{a},i}$, and $G_{\text{s},ij}$ for 10^4, 10^5, 10^6, 10^7, and 10^8 stochastic rays, normalized with respect to the reference solution for 10^9 rays, $\epsilon = \|\overline{y} - \overline{y}_{\text{ref}}\|_2 / \|\overline{y}_{\text{ref}}\|_2$, with $\overline{y}$ being the computed vector of the corresponding cumulative distribution function for the corresponding number of rays. The results are shown for the selected functions in figure 8.15.b.

ϵ decrease exponentially with N_{ray}. The maximum $\epsilon = 1.3$, obtained for $F_{\mu_{\text{in}},2}$ using $N_{\text{ray}} = 10^4$, decreases to $1.2 \cdot 10^{-2}$ using $N_{\text{ray}} = 10^8$. ϵ is relatively high for $F_{\mu_{\text{in}},2}$ because of the large values of the optical thickness inside the particles that leads to a small number of rays reaching particle boundaries and, consequently, high uncertainties in $F_{\mu_{\text{in}},2}$.

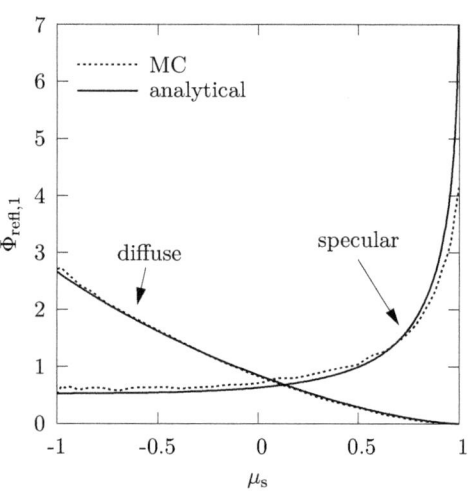

Figure 8.16: MC and analytically calculated phase functions of a particle cloud for $f_v = 1.6 \cdot 10^{-3}$, $d = 2$ µm, $n = 1.64$, and $k = 2.6 \cdot 10^{-5}$ at $\lambda = 1$ µm, $\rho_{\text{r,d}} = 0.866$ and for specularly and diffusively reflecting particles.

The MC algorithm was used to calculate the radiative characteristics of an independently scattering particle cloud of large opaque spheres for $f_v = 1.6 \cdot 10^{-3}$, $d = 2$ µm, $n = 1.64$, and $k = 2.6 \cdot 10^{-5}$ at $\lambda = 1$ µm. The scattering phase functions for diffusely reflecting and specularly reflecting particle surfaces are plotted in figure 8.16. Also shown are the corresponding phase functions obtained analytically [141]:

$$\Phi_{\text{refl},1,\text{d,an}}(\mu_s) = \frac{8}{3\pi}\left(\sqrt{1-\mu_s} - \mu_s \arccos(\mu_s)\right), \qquad (8.2)$$

$$\Phi_{\text{refl},1,\text{sp,an}}(\mu_s) = \frac{\rho_{\text{r,sp}}\left(1/2\left(\pi - \arccos(\mu_s)\right)\right)}{2\int_0^1 \rho_{\text{r,sp}}(\mu_{\text{in}})\mu_{\text{in}}d\mu_{\text{in}}}. \qquad (8.3)$$

The scattering and extinction coefficients can be calculated analytically [141] as follows:

$$\sigma_{\text{s,an}} = \frac{2}{3}\frac{f_v}{d}Q_s, \qquad (8.4)$$

8.3. Radiative Properties

$$\beta_{an} = \frac{2}{3}\frac{f_v}{d}Q_e, \qquad (8.5)$$

where $Q_{s,d} = \rho_d$, $Q_{s,sp} = \rho_{sp}$, and $Q_e = 1$ for large spheres without diffraction [141]. The MC-computed scattering coefficients for diffuse and specular particles, and the extinction coefficients are 1071 m^{-1}, 126 m^{-1}, and 1237 m^{-1}, respectively; the analytically calculated coefficients are 1032 m^{-1}, 121 m^{-1}, and 1200 m^{-1}, respectively. Furthermore, the code was validated for IOOS and IOTS [164].

The anisotropy of the sample was examined by subdividing the volume domain in eight subvolumes and calculating β_i along each coordinate axis for each subvolume petrasch2008, assuming opaque particles. The results are presented in figure 8.17; they indicate the negligible role of the sample's anisotropy.

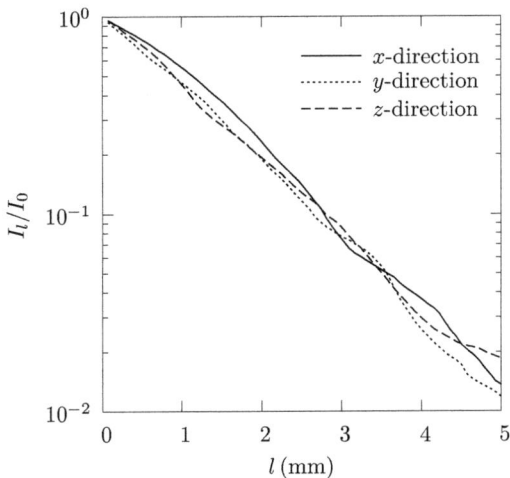

Figure 8.17: Normalized mean intensity along three orthogonal directions as a function of sample length.

The MC algorithm was further used to compute the radiative characteristics of a cubic ($\varepsilon = 0.48$) and orthorhombic ($\varepsilon = 0.40$) packed bed of glass spheres ($n = 1.5$, $k = 0$ at $\lambda = 0.6328$ µm) with a diameter of 1.269 mm. Since the glass particles are transparent at $\lambda = 0.6328$ µm, $I_1(\hat{s}) = I_2(\hat{s})$ [238] and $\beta_{MC} = \beta_1 \varepsilon + \beta_2 (1 - \varepsilon)$. β_{MC} was calculated to be 1775 m^{-1} and 1952 m^{-1} for the cubic and orthorhombic bed configurations, respectively. The experimentally determined value for a randomly packed bed ($\varepsilon = 0.42$) was $\beta_{ex} = 2326 \pm 416$ m^{-1} [98].

8.4 Conclusions

A computational technique has been developed in order to study the thermal radiative properties of two-phase media containing densely packed large non-spherical semitransparent particles. The 3D digital geometry of a packed bed of $CaCO_3$ particles was obtained by employing CT and was used to directly determine medium morphological characteristics such as porosity, specific surface, pore- and particle-size distributions, and the REV for the continuum domain. The collision-based MC method was applied to calculate the PDFs of the attenuation path length and the direction of incidence at the fluid-solid interface for the fluid and solid phases. The RDFs were then used to obtain the radiative properties of the two-phase medium.

Spectral extinction coefficients were found to be independent of the reflectivity type (specular/diffuse) of the fluid-solid interface. In contrast, they were found to depend strongly on the packed-bed geometry and the internal extinction coefficient $\kappa_{d,2}+\sigma_{d,s,2}$, which increases with wavelength due to the increasing complex refractive index. The scattering coefficients associated with refraction and reflection at the fluid-solid interface $\sigma_{s,ij}$ $(i \neq j)$ and $\sigma_{s,\text{refl},i}$, respectively, depend on the morphology and the reflective characteristics of the fluid-solid interface and on the internal radiative properties of the solid phase. $\sigma_{s,ij}$ and $\sigma_{s,\text{refl},i}$ behave complementarily. Scattering functions related to the refraction phenomenon Φ_{ij} $(i \neq j)$ are restricted by total reflection. $\Phi_{\text{refl},i}$ strongly depends on wavelength for specularly reflecting interfaces, while the opposite is true for diffuse-reflecting interfaces. Forward scattering was found to be predominant for specular-reflecting particles, while backscattering is typical for diffusely reflecting particles. Large internal extinction coefficients with isotropic internal scattering lead to less pronounced directional behavior of Φ_{ii}. Directional determination of the extinction coefficient elucidated the negligible anisotropy of the sample. The MC algorithm was validated by computing the radiative properties of semitransparent porous media reported in the literature.

The CT-based methodology's strength is its utilization of the exact 3D geometry of porous media in the limit of geometric optics with negligible diffraction. Once the accurate material's spectral bulk radiative properties and interface are known, the macroscale radiative properties can be accurately determined.

The methodology presented here can be applied generally for the determination of the radiative properties of porous materials with a complex geometry, e.g., radiation characteristic of snow samples (see next chapter), especially for low porosity media where measurements are difficult to carry out.

Chapter 9
Characterization of snow layers[1]

The theory and methodology for the radiative characterization complex multi-phase media, described in chapters 2 and 3, is applicable in multiple fields of engineering and science, as shown in this chapter for a multi-phase medium relevant in environmental science. The multi-phase medium investgated is snow, a packed bed of ice particles in a void phase composed of air and water vapor.

A multi-scale approach is chosen in this chapter: (i) tomography-based discrete-scale simulations are conduced to obtain the effective radiative properties of snow, namely: the extinction coefficients, the scattering coefficients and the scattering phase functions; (ii) the previously determined effective properties are used to solve the spatially averaged RTEs to determine overall reflectance, transmittance and absorptance of a snow slab (continuum model). The results of the model are validated through experimentally measured transmittance. Additionally, they are compared to models which assume simplified snow morphologies, allowing for conclusions on validity and accuracy of these simplified models.

The presence of soot and other impurities (ash, soil, bubbles) in snow can reduce the snow albedo and hence affect the radiative behaviour of the snow-pack [229, 155]. The tomography-based multi-scale approach allows for a straightforward quantification of these effects, which is shown in the last part of this chapter.

Radiative characterization of snow is of importance in environmental science, especially for energy and water balances [129, 130]. Climate models, modeling surface-atmosphere interactions, rely on the accurate prediction of the surface albedo of the snow cover [47, 182]. Remote sensing interpretation, crucial for the

---

[1]Material from this chapter has been submitted for publication: S. Haussener, M. Gergely, M. Schneebeli, and A. Steinfeld. Determination of the macroscopic optical properties of snow based on exact morphology and direct pore-level heat transfer modeling. *Journal of Geophysical Research*, submitted, 2012. [79]

determination of heat and mass transfer parameters in snow and consequently for avalanche prediction and energy balance, as well as hydrological and circulation models, strongly depends on the accurate radiative characterization of snow composed of different snow types [56, 151, 133].

Snow is a packed bed of semitransparent ice particles and semitransparent air and water vapor void phase. Models that are generally used to radiatively characterize snow are based on diluted spherical ice particles in air. These models determine the radiative properties by solving the problem of an electromagnetic wave incident on a spherical particle accompanied by an abruptly changing refractive index, corresponding to solving the Maxwell equations around a spherical particle [234, 227, 199, 16]. Nevertheless, these models reduce accuracy of the calculated radiative properties if applied to snow since they neither account for the complex ice grain morphology, which influences radiative properties [193, 168], nor for dependent scattering effects present for densely packed multi-phase media. A first approach to account for the complex morphology of snow has been introduced in CT-based radiation studies [99, 9]. Macroscopic optical properties, such as reflectance or transmittance of snow layers, have been calculated. But the models proposed do not allow to calculate the volumetric radiative properties such as extinction coefficient or scattering phase function, which are crucial for detailed snow cover models.

9.1 Morphological characterization

9.1.1 Computed tomography

Five different characteristic natural snow samples are used in the analysis, specified in table 9.1. They cover the grain shape classifications DFdc, RGsr/DFdc, RGsr, DHcp, and MFcl of the International Classification for Seasonal Snow on the Ground (ICSSG) [54]. They have been used in previous studies on snow's specific surface area [103]. Three types of snow called 'decomposing snow' (ds), 'metamorphosed I' (mI) and 'metamorphosed II' (mII) were prepared by sieving fresh snow into boxes after precipitation. The boxes were stored at different temperatures, allowing for isothermal metamorphism at different rates. Two additional snow samples were collected in the field: 'depth hoar' (dh) and 'wet snow' (ws). The dh snow was collected in blocks, while ws was sieved into boxes and soaked with ice water.

A modified Scanco µCT 80 desktop X-ray CT setup with a microfocus X-ray source emitting a polychromatic spectrum is used to scan the samples. The acceleration voltage of the generator is 45 keV. The sample is scanned at 1000

9.1. Morphological characterization

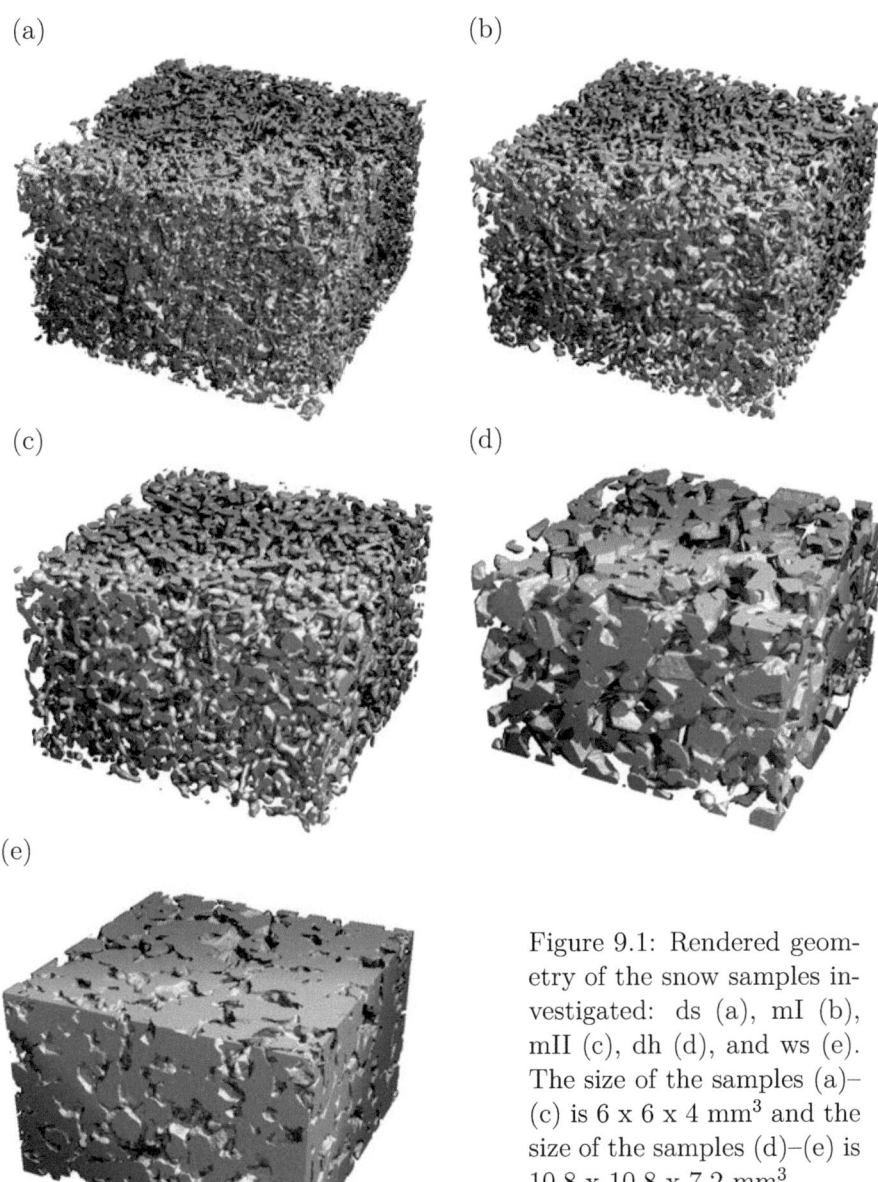

Figure 9.1: Rendered geometry of the snow samples investigated: ds (a), mI (b), mII (c), dh (d), and ws (e). The size of the samples (a)–(c) is 6 x 6 x 4 mm^3 and the size of the samples (d)–(e) is 10.8 x 10.8 x 7.2 mm^3.

Table 9.1: Grain shape classification (ICSSG) [54], measured density [103], preparation method, and voxel size of the CT scans of the characteristic snow samples.

Sample	ICSSG	ρ_{ex} (g/cm^3)	Preparation	Voxel size (µm)
ds	DFdc	0.11 ± 0.01	fresh snow for 8 days at -50 °C	10
mI	RGsr/DFdc	0.15 ± 0.01	fresh snow for 14 days at -17 °C	10
mII	RGsr	0.19 ± 0.05	fresh snow for 17 days at -3 °C	10
dh	DHcp	0.31 ± 0.02	snow from field	18
ws	MFcl	0.54 ± 0.03	field snow, soaked with water	18

angles (180°). Each measurement is an average of two scans with exposure times of 0.25 s. The resulting voxel sizes are 10 µm for the ds, mI and mII snow samples and 18 µm for the dh and ws snow samples. The CT data is filtered with a median and Gaussian filter (each 3 x 3 x 3) and subsequently segmented by mode method. For the ds snow sample the threshold for segmentation was determined by fitting two Gaussian curves to the grey value histogram and by calculating their intersection. This allows minimizing the amount of spurious voxels [103]. Figure 9.1 shows 3D rendered samples of the five characteristic snow samples.

9.1.2 Porosity, specific surface area and REV

The two-point correlation function is used to calculate porosity, ε, and specific surface area, A_0, of the characteristic snow samples. The specific surface area is defined as air-ice phase boundary surface per snow volume. Good agreement of porosity and specific surface to earlier measured values [103] is observed. REV is determined based on porosity calculations for cubic subvolume sizes at 20 random locations in the sample. Table 9.2 depicts the calculated ε, A_0 and edge lengths of the REV, l_{REV}, for the five snow samples. For the determination of l_{REV} a tolerance band in ε of ±0.05 is assumed.

9.1.3 Pore- and particle-size distributions

The mathematical morphology operation opening with a spherical structuring element is used to calculate the pore- and particle-size distribution, i.e. the size distribution defined by the sphere that fits completely within the pore or particle

9.1. Morphological characterization

Table 9.2: Calculated and measured [103] porosity and specific surface area, respectively, and REV length of the five characteristic snow samples.

Sample	ε	ε_{ex}	A_0 (m^{-1})	$A_{0,\text{ex}}$ (m^{-1})	$l_{\text{REV},\gamma=0.05}$ (mm)
ds	0.854	0.88 ± 0.01	8178	6776 ± 694	0.63
mI	0.845	0.84 ± 0.01	6450	5190 ± 383	1.27
mII	0.805	0.79 ± 0.05	5488	5130 ± 1473	1.37
dh	0.670	0.66 ± 0.02	2777	2883 ± 242	3.33
ws	0.384	0.40 ± 0.03	3016	2646 ± 219	3.93

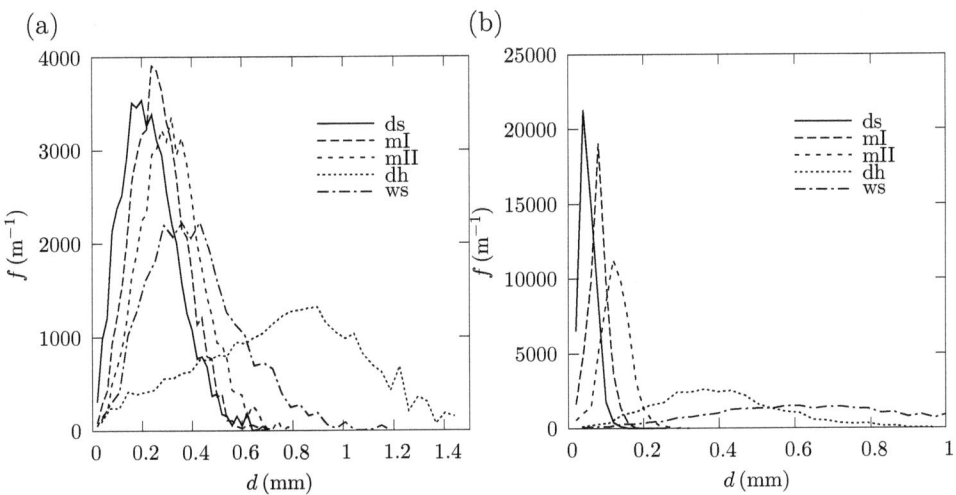

Figure 9.2: Opening size distribution functions of the pores (a), and the particles (b) for the five characteristic snow samples.

space, respectively. Figure 9.2 shows the opening distribution functions of the pore and particle sizes for the five snow samples. The mean, mode, median, and hydraulic diameters are listed in table 9.3. Thus, the assumption of geometrical optics is valid for $\lambda < 126$ µm.

Table 9.3: Calculated mean, mode, median and hydraulic pore and particle diameter of the five characteristic snow samples.

	ds		mI		mII		dh		ws	
	pore	particle	pore	part.	pore	part.	pore	part.	pore	part.
d_m	0.24	0.05	0.27	0.08	0.32	0.13	0.75	0.40	0.41	0.66
d_mode	0.20	0.04	0.24	0.08	0.32	0.12	0.90	0.36	0.43	0.61
d_median	0.27	0.07	0.28	0.10	0.33	0.15	0.81	0.42	0.43	0.70
d_h	0.42	0.07	0.52	0.10	0.59	0.14	0.97	0.47	0.51	0.82

9.2 Radiative characterization

9.2.1 Single-phase internal radiative properties

The fluid phase is assumed to be transparent, i.e. its internal absorption and scattering coefficients, $\sigma_{d,s,1}$, $\kappa_{d,1}$, respectively, are equal to zero, and its refractive index is equal to 1. The bulk properties of (pure) ice are determined based on the complex refractive index of ice (shown figure 9.3) determined in [228]. Internal scattering is assumed to be 0 and the absorption coefficient is calculated by the imaginary part of the refractive index, $\kappa_{d,2} = 4\pi k/\lambda$, according to the electromagnetic theory [20, 141]. The directional-hemispherical reflectivities at the specular fluid-solid interface for radiation incident from the fluid phase and for radiation incident from the solid phase, respectively, are calculated using Fresnel's equations accounting for absorption in ice [20, 141].

9.2.2 Two-phase medium radiative properties

Scattering and extinction coefficients

The scattering coefficients of the five characteristic snow samples are shown in figure 9.4 as a function of wavelength, for the fluid phase (figures 9.4.a and 9.4.b) and the solid phase (figures 9.4.c and 9.4.d). Obviously, for any phase, $\sigma_{s,ij}$ is complementary to $\sigma_{s,\mathrm{refl},i}$ since it refers to the transmitted portions of radiation across the interface. $\sigma_{s,\mathrm{refl},2}$ and $\sigma_{s,21}$ are additionally influenced by the presence of the total reflection phenomenon in the particle.

The extinction coefficients β_i are shown in figure 9.5. β_1 is independent of λ as it is a function of the interface geometry only. β_2 increases with λ, qualitatively following the λ-dependence of $\kappa_{d,2}$. For $\lambda > 2.7$ µm, internal scattering strongly increases and leads to the black body like behavior of ice in the near-IR spectral

9.2. Radiative characterization

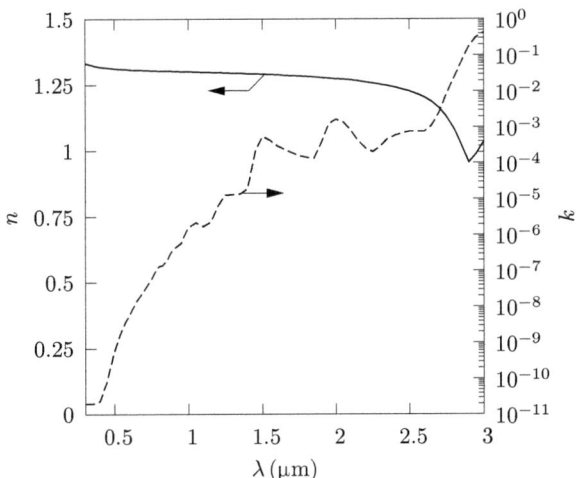

Figure 9.3: Real (solid line) and imaginary (dashed line) part of the complex refractive index of ice for 0.3 µm $< \lambda <$ 3 µm [228].

region.

Scattering phase functions

Figure 9.6 shows the scattering phase functions for the five characteristic snow samples as a function of the cosine of the scattering angle, μ_s. The sharp increase in $\Phi_{\text{refl},2}$ can be explained by the total reflection phenomena leading to an increased fraction of forward scattering. The scattering phase functions for the five characteristic snow sample behave nearly identical and no significant wavelength dependence is observed. This finding is consistent with the low sensitivity of the two-phase medium scattering functions on morphology [211, 78].

9.2.3 Continuum properties

Methodology

In order to obtain the macroscopic optical properties (overall reflectance, transmittance and absorptance) of the five characteristic snow samples, a spatially averaged model (continuum domain model) is developed. The previously determined two-phase medium properties account for the complex geometry and the single phase internal radiative properties. A snow slab of thickness l_{slab} is

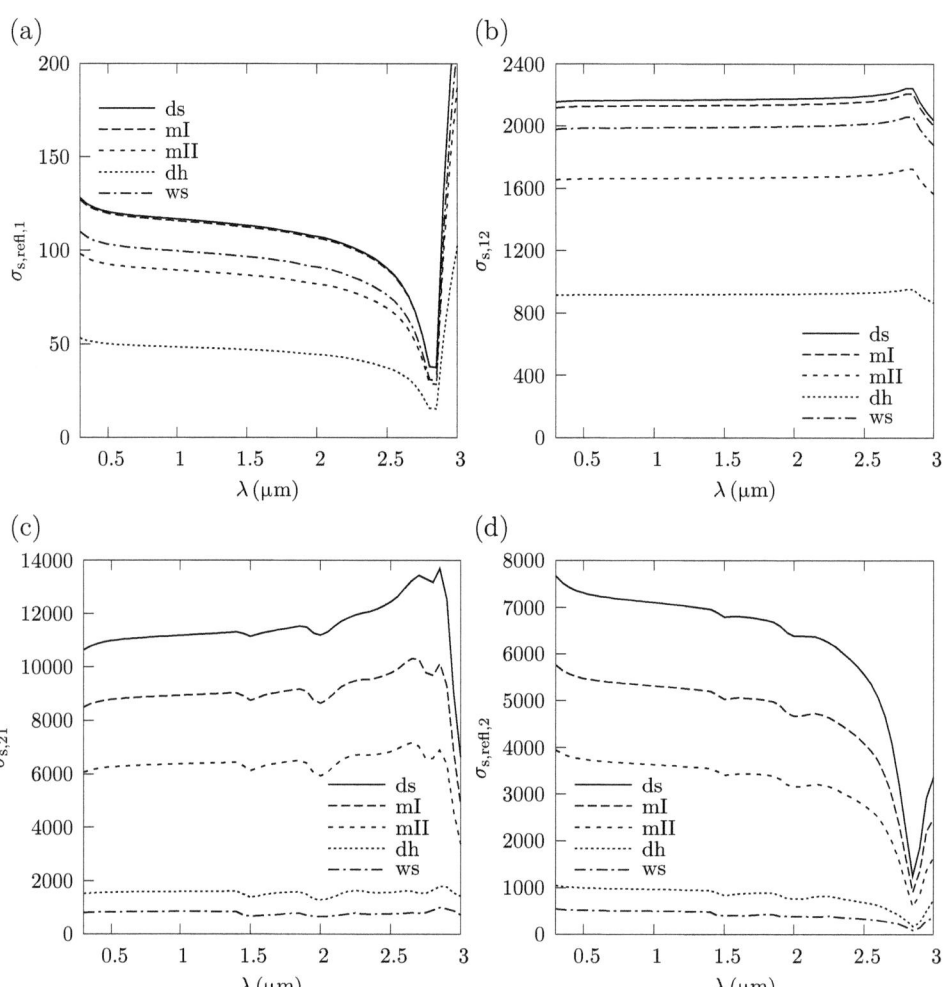

Figure 9.4: Spectral scattering coefficients of the five characteristic snow samples for the fluid phase (a,b) and the solid phase (c,d).

9.2. Radiative characterization

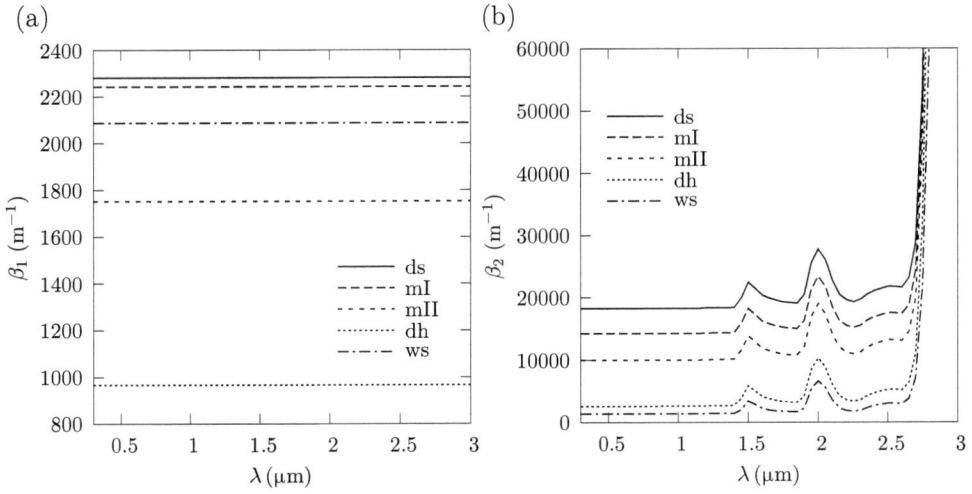

Figure 9.5: Spectral extinction coefficients of the five characteristic snow samples: fluid phase (a) and solid phase (b).

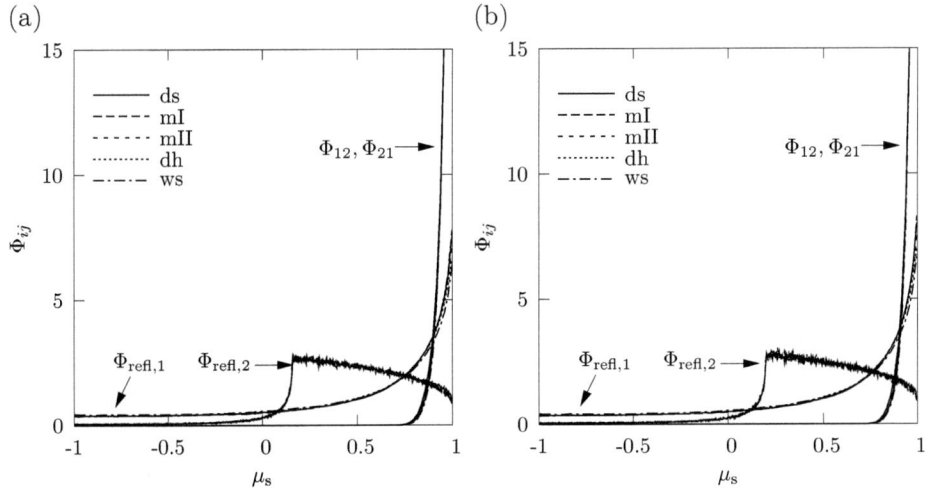

Figure 9.6: Scattering phase functions of the five characteristic snow samples at $\lambda = 0.5$ µm (a) and at $\lambda = 1.5$ µm (b).

exposed to (i) a collimated incident radiation flux or (ii) a diffuse incident radiation flux at $z = 0$. The situation is depicted in figure 9.7. The lateral edges of the slab are assumed infinitely large, equal to a periodic boundary. The inlet

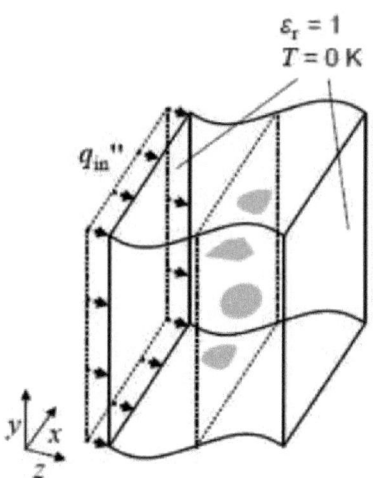

Figure 9.7: Schematic of the snow slab sample used for the continuum domain calculations. The inlet radiative heat flux is depicted by q''_{in}, the black body surrounding at the inlet and outlet is given by $T = 0$ K and $\varepsilon_{\text{r}} = 1$, respectively. The lateral walls are infinitesimally large.

and outlet faces are exposed to nonparticipating black surrounding, therefore $T = 0$ K and emissivity $\varepsilon_{\text{r}} = 1$. Path-length based MC technique is used to solve the two coupled averaged RTEs. Overall reflectance, R, transmittance, Tr, and absorptance, A, of the slab with thickness l_{slab} are defined by Eqs. (9.1)–(9.3). $\overline{q_i^j}$ describes the radiative flux averaged over the cross sectional area in forward or backward direction (j: $+, -$), and in the void or solid phase (i: 1, 2).

$$R = \frac{\overline{q_1^-}(z=0) + \overline{q_2^-}(z=0)}{q_{\text{in}}}, \qquad (9.1)$$

$$Tr = \frac{\overline{q_1^+}(z=l_{\text{slab}}) + \overline{q_2^+}(z=l_{\text{slab}})}{q_{\text{in}}}, \qquad (9.2)$$

$$A = 1 - R - Tr. \qquad (9.3)$$

Radiative properties

The calculated R, Tr and A of the five characteristic snow samples, spectrally resolved, are shown in figure 9.8 for $l_{\text{slab}} = 4$ cm for collimated incidence radiative flux. Reflectance and absorptance closely follow the λ-dependence of $\kappa_{\text{d},2}$. The results are qualitatively in agreement with the radiative properties of snow with

9.2. Radiative characterization

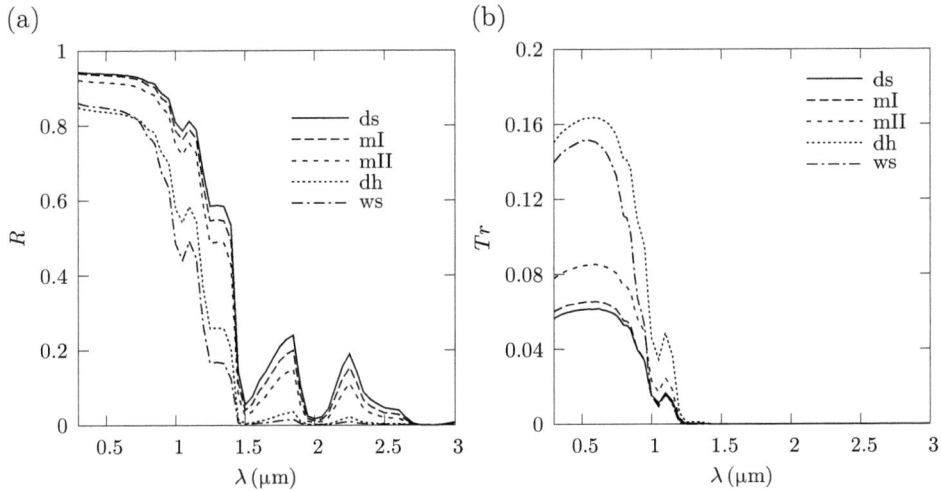

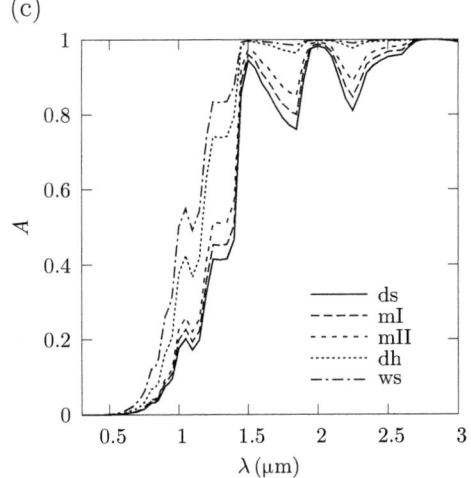

Figure 9.8: Spectral dependence of the overall reflectance (a), transmittance (b), and absorptance (c) of a slab with l_{slab} = 4 cm and composed of the five characteristic snow types and irradiated by collimated radiative flux.

varying spherical grain radius determined in [234]. Comparison between R and Tr calculated for collimated or diffuse incident radiative flux are depicted in figure 9.9. R for a diffusely irradiate boundary increases by 1% (at $\lambda = 0.5$ µm) up to 13% (at $\lambda = 1.4$ µm) for ds, mI and mII. For dh and ws R increases even by 40% (at $\lambda = 1.4$ µm). Tr for a diffusely irradiated boundary decreases by 27% (at $\lambda = 0.5$ µm) and by 50% (at $\lambda = 1.4$ µm) for all five snow types. A decreases for a diffusely irradiated boundary.

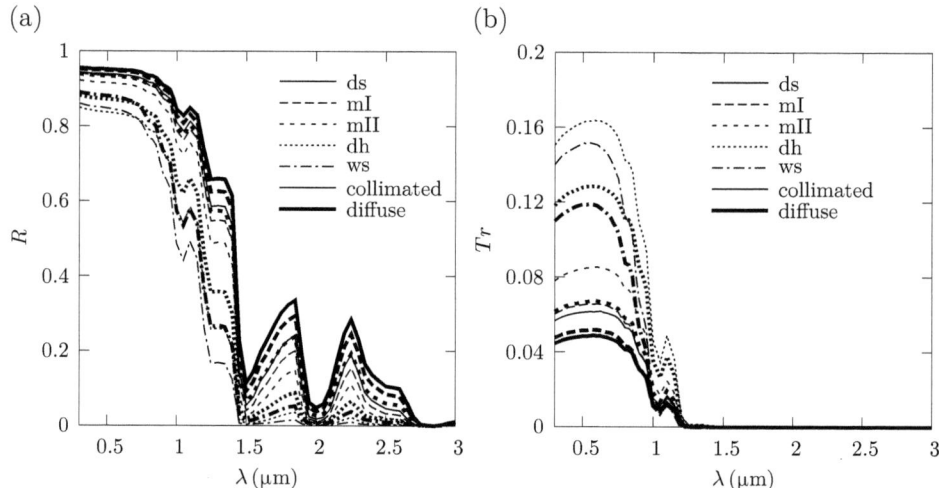

Figure 9.9: Comparison of the spectral dependence of the overall reflectance (a), and transmittance (b) of a slab with $l_{\text{slab}} = 4$ cm, composed of the five characteristic snow types and irradiated by collimated (thin solid line) and diffuse (thick solid line) radiative flux.

Transmittance as a function of sample thickness is depicted in figure 9.10 for ds and ws snow types. The expected exponential decrease in Tr with increasing l_{slab} is observed. Figure 9.11 shows transmittance as a function of snow types for 2 cm and 8 cm thickness, respectively. The behavior qualitatively follows the mean pore sizes of the five characteristic samples (see table 9.3).

Discrete-scale radiative heat transfer

In addition to the determination of the macroscopic optical properties based on the exact snow morphology, the proposed CT-based multi-scale methodology allows for an in-depth heat transfer analysis on the exact microstructure based on direct numerical simulations at the pore-level scale. This information is of inter-

9.2. Radiative characterization

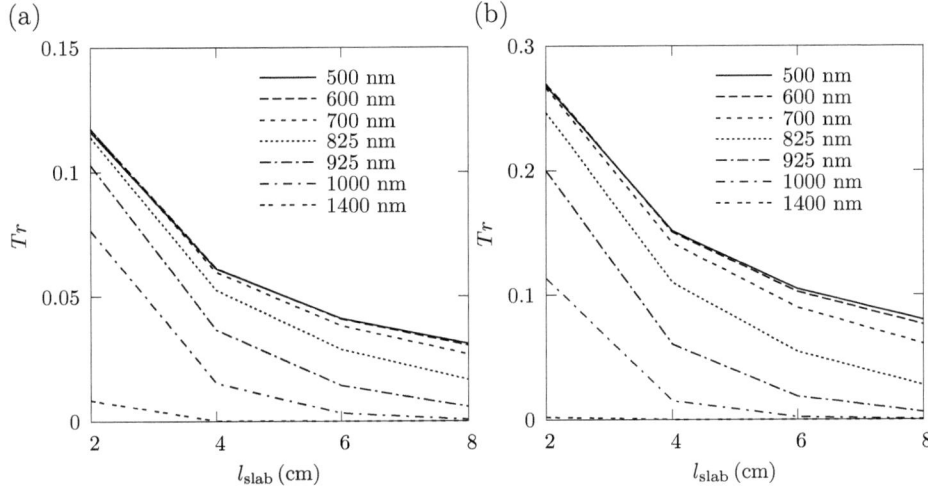

Figure 9.10: Transmittance in function of the sample's thickness for different wavelengths for two characteristic snow samples: ds snow (a) and ws snow (b).

est for energy balance between ground and atmosphere in snow covered regions where the interplay of the different heat transfer modes needs to be analyzed at the pore-level scale, e.g. when snow cover gets unstable when irradiated by different direction and intensity of solar irradiation. Exemplary, the absorbed radiation in the xy-plane at $z = 0.2$ mm within a 6 x 6 x 4 mm^3 mII type snow sample irradiated with diffuse radiation at $z = 0$ and black surrounding for the other five walls is depicted in figure 9.12 for $\lambda = 1$ and 2.5 µm. Radiation is predominantly absorbed near the surface of the ice particle. At $\lambda = 2.5$ µm, internal absorption within ice is two orders of magnitude larger (see figure 9.3) and, therefore, more radiative energy is absorbed in the ice matrix. In addition, the effective (volumetric) radiative properties determined allow for a statistical and spatial averaged investigation of radiative heat transfer at the pore-level scale, shown in figure 9.12.d for the mII snow type at $\lambda = 1$ µm, with significant savings in computational expenses [161]. The volumetric absorption within the mII and ws type snow samples is depicted in figure 9.13 for $\lambda = 2.5$ µm.

Experimental validation

Experimental validation of the calculated results and consequently the CT-based multi-scale methodology proposed is presented in the following. Experimentally

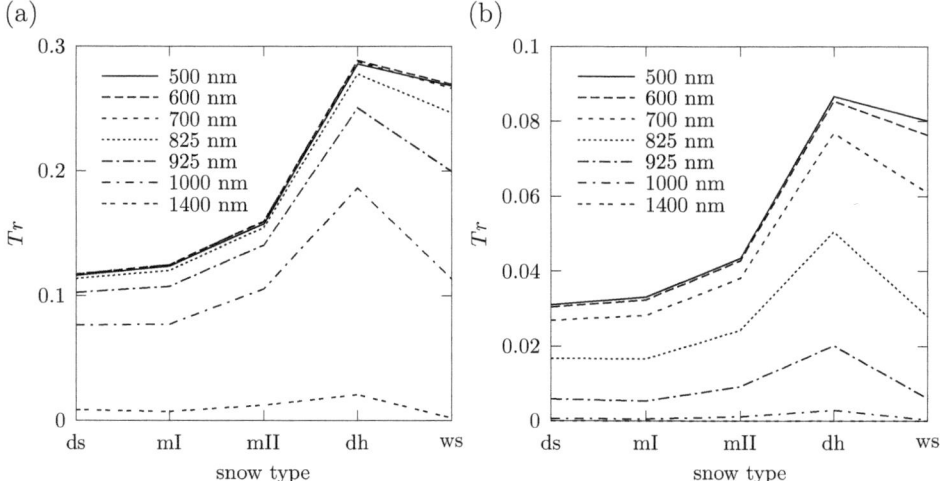

Figure 9.11: Transmittance in function of the five characteristic snow samples for varying λ and for $l_{\text{slab}} = 2$ cm (a) and 8 cm (b).

estimated transmittances of different snow samples are determined with a self-built integrating sphere setup [64]. A 60 W halogen lamp is used as radiation source. The sphere is made of styrofoam with an interior coating of aluminium foil, spray-painted in flat white. It diffusely illuminates the snow sample of thickness l_{slab}. The radiation transmitted through the snow sample is measured by an Analytical Spectral Device (FieldSpec Pro) in the spectral range of 350 to 1050 nm.

Two sets of experiments are performed. In the first set the transmittance measurements of a snow sample collected in the field are done at two wavelengths (830 and 928 nm) for 4 different sample thicknesses ($l_{\text{slab}} = 2.5, 4, 6,$ and 8 cm). Measured ε of the sample is in the range of 0.33 to 0.52 and measured A_0 is in the range of 3300 to 3900 m^{-1}. The experimentally estimated transmittance is compared to the transmittance calculated for the dh and ws samples, which show similar ε and A_0. The results are depicted in figure 9.14.a. The error bars account for the variation in sample thickness and an assumed uncertainty in transmittance measurement of 10%. The slightly lower calculated transmittance at lower wavelengths is explained by the generally lower porosity of the dh and ws samples.

9.2. Radiative characterization

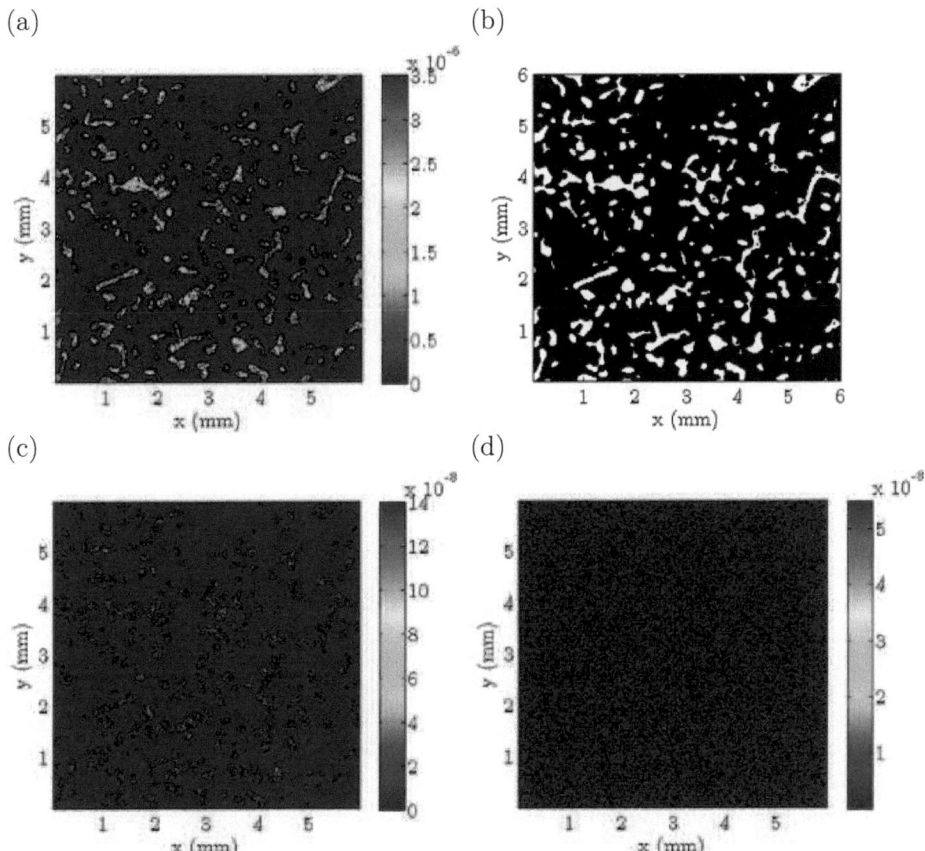

Figure 9.12: Fraction of incident energy absorbed at the pore-level scale in the xy-plane ($z = 0.2$ mm) of the mII type snow for $\lambda = 1$ µm (c) and 2.5 µm (a). The corresponding snow structure is shown in (b), where black denotes void/air phase and white the ice particle. The statistically and spatial averaged absorption behavior in mII snow type for $\lambda = 1$ µm at $z = 0.2$ mm is depicted in (d).

In a second set of experiments, samples of three additional snow types are investigated: 'refrozen wet natural snow' (ns, small melt forms) with mean l_{slab} = 4.5 cm, 'stomped ns' (small rounded grain) with mean l_{slab} = 4.85 cm and 'stomped snowmaker snow' (ss, fine grains) with mean l_{slab} = 4.75 cm. After each the measurement, three subsamples of each snow type are scanned by CT and used for transmittance calculations with the proposed CT-based multi-scale

162　　　　　　　　　　　　　Chapter 9. Characterization of snow layers

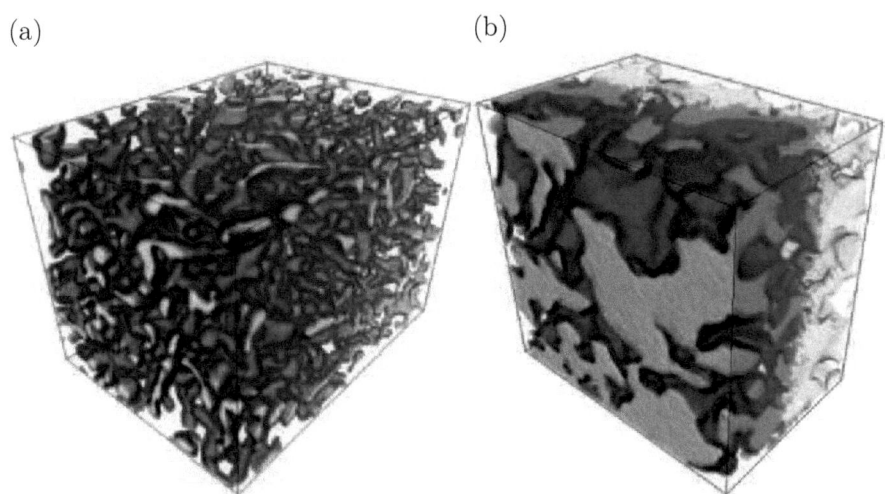

Figure 9.13: Fraction of incident energy absorbed at the pore-level scale in a 3 x 3 x 4 mm^3 mII type (a) and a 5.4 x 5.4 x 7.2 mm^3 ws type (b) snow sample at $\lambda = 2.5$ µm. The samples are diffusely irradiated from the front.

methodology. The results of the comparison are depicted in figure 9.14.b. The error bars in the calculations account for the morphological variation between the three subsamples and the variation in sample thickness. Results and calculations are in good agreement, except for the refrozen wet ns sample where acceptable agreement is obtained. The discrepancies are attributed to inhomogeneities in the snow sample and measurement uncertainties, especially at these low transmittances.

Comparison to simplified morphologies

To investigate the influence of the incorporation of the exact snow sample morphology on radiative properties e.g. by means of CT, on the radiative properties the previously obtained results are compared to radiative characterization of idealized snow types. An artificially generated regularly structured snow sample, a packed bed of identical overlapping semitransparent spheres (IOSS), is used for comparison. Porosity, ε_{IOSS}, and specific surface area, $A_{0,IOSS}$, of IOSS are given by

$$\varepsilon_{IOSS} = \exp(nd^3 \frac{\pi}{6}), \qquad (9.4)$$

9.2. Radiative characterization

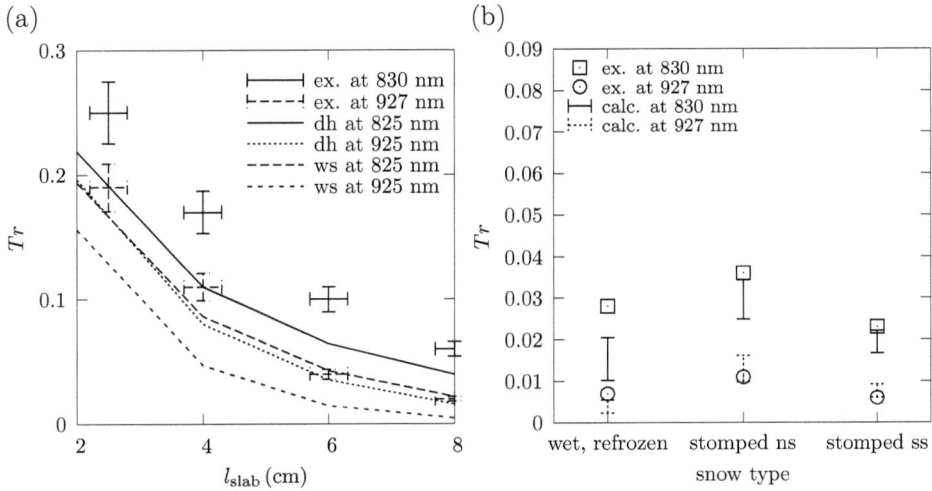

Figure 9.14: Transmittance as a function of sample thickness calculated for a ws snow sample at 825 and 925 nm (solid and dashed lines) compared to experimentally determined transmittance of a similar snow type at 830 and 927 nm (solid and dashed lines) (a); and experimentally determined transmittance of wet refrozen snow, stomped ns, and stomped ss with mean thicknesses of 4.5, 4.85 and 4.75 cm, respectively (b).

$$A_{0,\text{IOSS}} = -\frac{6\varepsilon_{\text{IOSS}} \ln(\varepsilon_{\text{IOSS}})}{d}. \tag{9.5}$$

n describes the number of spheres per volume and d the diameter of the identical spheres. Three types of IOSS samples are generated: (i) with d equals to the corresponding d_m (given in tables 9.2 and 9.3) and $\varepsilon_{\text{IOSS}}$ equals to the corresponding ε of the snow type; (ii) with $A_{0,\text{IOSS}}$ equals to the corresponding A_0 of the snow type and $\varepsilon_{\text{IOSS}} = \varepsilon$; and (iii) with $d = d_\text{m}$ and $A_{0,\text{IOSS}} = A_0$.

The comparison of R for snow slabs of ds and ws snow types calculated based on the CT morphology or the 3 types of IOSS morphologies with each having two common parameters ($d = d_\text{m}$ and $\varepsilon_{\text{IOSS}} = \varepsilon$; or $A_{0,\text{IOSS}} = A_0$ and $\varepsilon_{\text{IOSS}} = \varepsilon$; or $d = d_\text{m}$ and $A_{0,\text{IOSS}} = A_0$) are shown in figure 9.15. Table 9.4 indicates the normalized 2-norms,

$$\epsilon = \frac{\|R_{\text{IOSS},\text{type}_i} - R_{\text{CT}}\|_2}{\|R_{\text{CT}}\|_2}, \tag{9.6}$$

and the relative difference,

$$r_\text{dif} = \frac{|R_{\text{IOSS},\text{type}_i} - R_{\text{CT}}|}{|R_{\text{CT}}|}, \tag{9.7}$$

of the calculated R based on the idealized morphologies and exact morphology.

Of the five characteristic snow types investigated, a slab composed of ds snow type is worst approximated by IOSS while a slab of mII snow is best approximated by IOSS. This is due to the fact that the grains of mII are closest to a sphere. The IOSS with $A_{0,\text{IOSS}} = A_0$ and $\varepsilon_{\text{IOSS}} = \varepsilon$ best approximates the morphology. Nevertheless, the relative difference of the calculated morphology still deviates by 6 (mII) to 24% (ws).

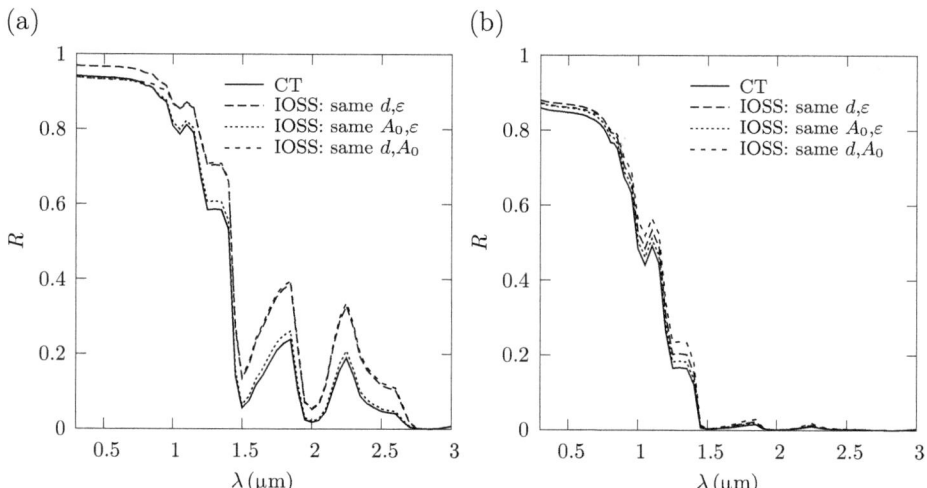

Figure 9.15: Comparison of R of a ds snow slab (a) and ws snow slab (b) of l_{slab} = 4 cm using either exact snow morphology (CT) or idealized morphology (3 types of IOSS).

9.2.4 Snow with soot impurities

To demonstrate the power of the CT-based multi-scale methodology proposed, the influence of impurities in snow on radiative behavior is analyzed. Snow impurities most often consist of dust, soil or soot. Additionally, bubbles in the ice grain may act as radiation scatter. Soot impurities are chosen for the following analysis due to its largest expected influence on snow albedo and, therefore, on climate and hydrological models [229, 155, 173]. Nevertheless, the same procedure is applicable for dust or soil impurities, or bubble inclusions.

9.2. Radiative characterization

Table 9.4: Normalized, 2-norm ξ, and relative difference at $\lambda = 1$ µm, $r_{\text{dif},\lambda=1}$, of the calculated R based on exact morphology and idealized geometry.

	CT vs. d_m/ε-IOSS		CT vs. A_0/ε-IOSS		CT vs. d_m/A_0-IOSS	
	ϵ	$r_{\text{dif},\lambda=1}$	ϵ	$r_{\text{dif},\lambda=1}$	ϵ	$r_{\text{dif},\lambda=1}$
ds	85.7	0.0770	12.3	0.0139	87.8	0.0734
mI	59.2	0.0653	13.2	0.0116	61.5	0.0595
mII	42.1	0.0601	6.4	0.0061	42.2	0.0569
dh	61.4	0.1252	11.4	0.0238	52.3	0.1288
ws	50.1	0.0876	25.4	0.0406	76.8	0.1536

Single phase properties of ice grains with soot impurities

Soot impurities are most likely randomly distributed within the ice grains [17]. Therefore the properties of the ice grains have to be adapted in order to adequately account for soot impurities. Measured soot impurities are in the range of 1 to 10 ppm weight fraction (corresponding to approximately 1 to 10 ppm f_v) [229]. Mie theory [16] is used to calculate the radiative behavior of an ice particle with soot impurities. The refractive index of soot [43, 27, 110, 17], shown in figure 9.16, together with an approximate diameter of (primary) soot particles of 10 to 100 nm [229, 17] are assumed for the calculations. It is assumed the soot particles are isotropically scattering ($\Phi_{d,2} = 1$). The radiative behavior of the ice/soot mixture is then determined by:

$$\kappa_{d,2,\text{ice}} = \frac{4\pi k_{\text{ice}}}{\lambda}, \tag{9.8}$$

$$\kappa_{d,2,\text{imp}} = \frac{3}{2}\frac{f_v}{d}Q_a, \tag{9.9}$$

$$\sigma_{d,s,2,\text{imp}} = \frac{3}{2}\frac{f_v}{d}Q_s, \tag{9.10}$$

$$\beta_{d,2} = (1 - f_v)\kappa_{d,2,\text{ice}} + \kappa_{d,2,\text{imp}} + \sigma_{d,s,2,\text{imp}}. \tag{9.11}$$

Q_a and Q_s are the absorption and scattering efficiencies calculated based on Mie theory. Since soot particles are probably agglomerated, additionally, an equivalent sphere approach [110] is used to account for the agglomerated soot particles. The diameter of the equivalent sphere of the agglomerate is determined by:

$$d_{\text{eq}} = \sqrt[3]{N_{\text{p/agg}}} d_p. \tag{9.12}$$

For the calculations it is assumed that an agglomerate consists of approximately 400 primary particles, $N_{\text{p/agg}} = 400$ [110], of 10 to 100 nm diameter.

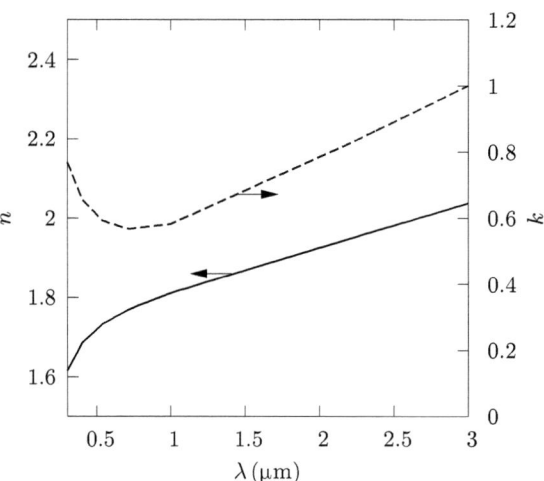

Figure 9.16: Real (solid line) and imaginary (dashed line) part of the complex refractive index of soot for 0.3 µm $< \lambda >$ 3 µm [27].

Direct pore-level radiative properties

The determination of the two-phase medium radiative properties of the snow sample of different snow types, containing soot impurities, is straightforward; it is described in chapter 2. There is no change in $\sigma_{s,ij}$ $(i \neq j)$ and $\sigma_{s,\mathrm{refl},i}$ compared to the pure snow sample since they solely depend on the morphology of the snow sample. The same is valid for $\Phi_{s,ij}$ $(i \neq j)$ and $\Phi_{s,\mathrm{refl},i}$. $\beta_{d,2}$ is changing due to the soot impurities, and consequently β_2, $\sigma_{s,22}$ and $\Phi_{s,22}$ are changing.

Continuum model

The continuum model, described in section 9.2.3, is used to calculate reflectance, transmittance and absorptance of snow slabs composed of the five characteristic snow types, containing soot impurities. The calculated spectral reflectance of a snow slab with $l_{\mathrm{slab}} = 4$ cm composed of ds snow or ws snow are shown in figure 9.17. A more pronounced influence on R is observed for the sample composed of larger ice grains. For small λ and $d_\mathrm{p} = 10$ nm and $f_v = 10^{-5}$ R is reduced up to 46% and 87% for the samples composed of ds and ws snow, respectively. For $\lambda > 1.4$ µm, no significant influence of soot impurities on R is observed. The influence of soot particle size (d_p) is less pronounced than the volume fraction of the soot particles (f_v). If the soot is modeled as agglomerates, the influence on R is less pronounced.

9.2. Radiative characterization

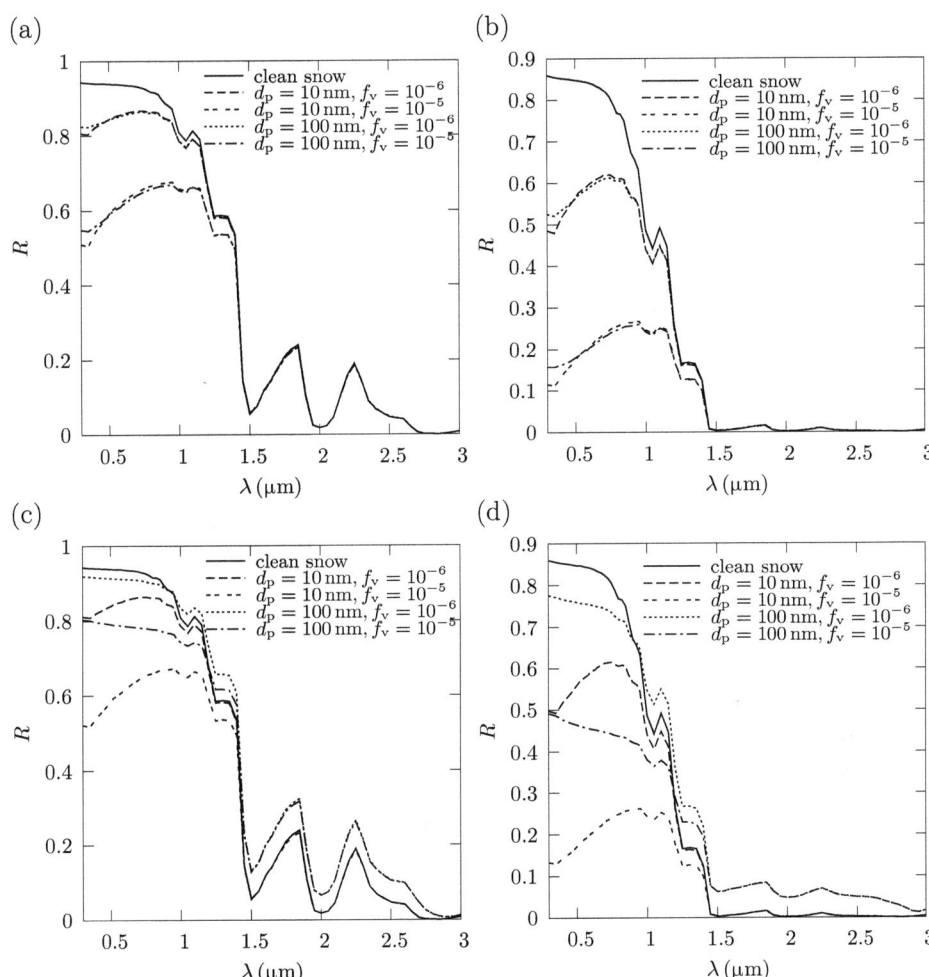

Figure 9.17: Reflectance of a snow slab of $l_{\text{slab}} = 4$ cm of ds snow type (a,c) and ws snow type (b,d) containing soot impurities, either modeled as single particles homogeneously distributed within the ice (a,b) or as agglomerates composed of the same size primary particles homogeneously distributed within the ice (c,d).

9.3 Conclusions

For the accurate determination of spectrally resolved (0.3 to 3 µm) radiative characteristics of different snow types, a tomography-based multi-scale methodology was proposed. As characteristic snow types, decomposed snow, two types of metamorphosed snow (composed of small and large rounded grains), depth hoar and wet snow were chosen. The complex morphology of the different snow types had been digitalized by CT and were used in direct discrete-scale numerical simulations for the determination of the morphological and effective radiative properties, namely: porosity, specific surface, pore- and particle-size distribution, edge length of the representative elementary volume, extinction coefficients, scattering coefficients and scattering phase functions. A continuum model was developed to determine overall reflectance, absorptance and transmittance of snow slabs composed of the different snow types.

The results obtained were in agreement with previously published radiative properties of snow calculated by alternative methods, based on simplified snow morphologies. In turn, the calculated properties were in agreement with the experimentally estimated transmittance of snow samples, obtained by a self-built integrating sphere setup [64]. The comparison between radiative properties of snow samples composed of the five different characteristic snow types obtained based on the exact snow morphology or simplified snow morphologies showed discrepancies in reflectance up to 24%.

The power of the CT-based multi-scale methodology developed was shown by additionally calculating radiative characteristics of snow slabs composed of different characteristic snow types containing soot impurities. Soot particles of 10 nm diameter and present in a volume fraction of 10^{-5} can reduced the reflectivity by 47 to 83%. This reduction potential of soot was decreased when modelling the soot impurities as agglomerates.

The CT-based, multi-scale methodology proves to be an accurate and powerful method to determine effective radiative properties of snow based on the exact and complex snow morphology. Radiative characteristics of snow composed of stratified slabs, each slab composed of another snow types, can be modeled straightforwardly with the proposed method.

Chapter 10
Summary and outlook

This book examines complex, chemically reacting multi-phase media in multi-scale systems used in solar energy applications. In the course of this book, the derivation of the volume-averaged radiative transfer equations and, consequently, the effective radiative properties of multi-phase media was achieved. The accurate determination of the morphological, effective heat and mass transfer properties, based on the exact discrete-scale geometry obtained by computed tomography, was achieved for (i) highly porous ceramic foams, (ii) anisotropic ceramic foams, (iii) packed beds of reacting material, and (iv) packed beds of semitransparent particles. A fifth medium relevant in the area of environmental sciences and climate modeling was investigated to show the wide applicability of the methodology: (v) layers of characteristic snow types. Experimental validation was carried out for the morphological and effective radiative properties. Preliminary studies on tailored ceramic foam design, adapted to the specific process needs, were conducted. Continuum models for the highly porous ceramic and the layers of characteristic snow were developed. The properties of the highly porous ceramic foam were incorporated in a continuum model of a solar reactor used for the evaporation and decomposition of sulfuric acid, the high-temperature step of sulfur-based cycles for the production of hydrogen. Comparison with temperature measurements of experiments obtained at the solar furnace of DLR in Cologne was done. The subsequent process optimization of the operational conditions, the reactor design, and the foam morphology was analyzed. The properties of the snow were incorporated in a continuum model of an irradiated snow slab and analyzed for the macroscopic optical properties. Comparison with transmission measurements of snow showed good agreement. Subsequent analysis of the influence of snow morphology and snow impurities on the macroscopic optical properties was carried out.

The thesis from which this book evolved was performed in the framework of the Hycycles and the SolRad projects. HycycleS is an ongoing project and aims at the qualification and enhancement of materials and components for key steps of sulfur-based thermochemical and thermochemical/electrochemical cycles for solar or nuclear hydrogen generation. Its final aim is to put to practice thermochemical water-splitting by improving the efficiency, stability and practicability and by lowering the costs. The task of ETH was to develop the detailed heat and mass transfer model in the porous foam of the solar evaporator/decomposer reactor and it subsequent optimization. The SolRad project aimed at the investigation of the fundamental heat and mass transfer phenomena in chemically reacting multi-phase media applied to the production of hydrogen via gasification of carbonaceous materials under concentrated solar radiation.

In chapter 2, an introduction to volume averaging theory is given. The volume-averaged radiative transfer equations (RTEs) were derived and used for the derivation of effective radiative properties in multi-phase media, assuming geometrical optics to be valid. The general approach was simplified for two-phase media composed of semitransparent phases, where the derivation showed that two averaged RTEs need to be solved, each requiring an extinction coefficient, two scattering coefficients and two scattering phase functions. It was shown that the properties solely depend on the discrete-scale geometry of the multi-phase material and its component's intrinsic radiative properties. Additionally, a brief description of the closure of the volume-averaged mass, species, momentum and energy equations was given.

The methodology based on discrete-scale numerical simulations is introduced in chapter 3. It is used for the investigation of the morphological, effective heat and mass transfer properties. The methodology was extended to the specific multi-phase media's requirements analyzed in this book. Radiative distribution functions, computed by collision-based Monte Carlo, was used for radiative characterization. Finite volume techniques were used for the solution of the discrete-scale mass, momentum, and energy equations, needed for the conductive, convective and mass transfer characterization. As the effective properties of multi-phase media solely depend on component's intrinsic properties and discrete-scale geometry, a computed-tomography (CT) based approach for obtaining an accurate discrete-scale geometry representation was used. The morphological characterization by CT, segmentation of the CT data and its subsequent use for the determination of porosity, specific surface area, pore- and particle-size distributions and representative elementary volume, are described in chapter 4.

The governing equations and the corresponding methodology for the solution of the equations were then applied to the characterization of four different

Chapter 10. Summary and outlook 171

multi-phase media used in four different solar applications and one multi-phase medium relevant in environmental science.

The sulfur-based water-splitting cycles are introduced in chapter 5. A solar reactor, hosting reticulate porous ceramic (RPC), was proposed for the evaporation/decomposition step. Therefore, RPC with nominal pore diameter of 1.27 mm was characterized for morphological, effective heat and mass transfer properties. The extinction coefficient was calculated to be 431 m^{-1} and was in quantitative agreement with its experimental estimate of 673 ± 30 m^{-1}. Discrepancies were assumed to be the cause of neglecting incoming scattering in the experimental methodology. The calculated scattering phase functions for the limiting cases of pure diffuse or specular surface reflection showed predominant back-scattering or nearly isotropic behavior, respectively. The effective conductivity converged to $0.022 \cdot k_s$ for $k_f \ll k_s$ and the Nu correlation converged to Nu = 6.8 for small Re numbers. Calculated permeability and Dupuit-Forchheimer coefficient compared well to values available in the literature for similar discrete-scale geometries. Tortousity showed a mean value of 1.07. The dispersion tensor for neglected molecular dispersion was determined as a function of Re. The effective properties were then incorporated in a continuum model of the solar reactor containing the foam. Experimental comparison by means of measured temperatures showed good agreement. The model was used for the optimization of the operational conditions, reactor design and foam morphology. Acid solution inflow and solar irradiation showed to be the most important parameters. For the current design a solar irradiation of 150 W and an acid solution inflow of 2 ml/min achieved highest energetic and chemical efficiencies of 73% and 45%, respectively.

A ceria-based water-splitting cycle is introduced in chapter 6. The non-stoichiometric ceria foam used in the solar reactor for its redox reactions showed structural anisotropy due to its processing (uniaxial pressing in z-direction and anisotropic primary particles). Therefore, morphological and effective transport properties of different ceria samples (with different pore-size distributions) were determined along the three principal directions. Pore squeezing in z-direction led to the enhanced extinction behavior in z-direction, consistent – also in the other two directions – with values available in literature for similar media. The calculated effective conductivities showed enhanced conductivity along the x- and y-directions due to a more parallel alignment of the structure with the heat flux in these directions. The determined Nu correlations showed enhanced convective heat exchange along the z-direction. Permeabilites decreased while Dupuit-Forchheimer coefficients increased for the z-direction because of an enhanced pressure gradient due to hindered fluid paths and considerable increase in tortuosity for the z-direction. Finally, preliminary studies on tailored foam design, adjusted to the specific process needs, were carried out, with the aim to analyze and optimize the artificially generated ceramic foams and their effective heat and mass transfer properties.

Solar-driven gasification of carbonaceous material (waste tire shreds) in a packed-bed configuration is introduced in chapter 7. An energy analysis of different electricity production processes using waste tire shreds as feedstock was conducted. Thus, its applicability and its CO_2 emission potential was shown. In the subsequent analysis, the accurate determination of morphology, effective heat and mass transfer properties dependent on reaction extent was achieved. In contrast to the previously investigated ceramic foams, the packed bed was a non-consolidated structure and its morphology considerably changed during the reaction. Experimentally determined porosity for different operational conditions increased with reaction extent, peaked at $X_C = 0.86$ and decreased again. Comparison to numerically calculated values showed good agreement for low X_C, while at large X_C discrepancies, associated to evolved nanopores not detectable with the CT techniques, were observed. Measurements of BET specific surface area supported this explanation. Measured particle-size distributions showed that during pyrolysis, where devolatilization takes place, the particle became highly porous without significant decrease in diameter. On the other side, particles shrank and broke apart during subsequent gasification resulting in substantially smaller particle sizes. Numerically calculated particle-size distribution were able to predict the increase in small particles with increasing X_C. Calculated extinction coefficients increased with reaction extent as particles shrank, broke apart and shortened the attenuation path length. Scattering phase functions for diffusely reflecting particles showed no significant influence on the reaction extent while the opposite was true for specularly reflecting particles. Effective conductivity, calculated by neglecting the particle-particle contact resistance, decreased with reaction extent because of the increase in porosity and the decrease in particle size. Nu correlations showed a decrease in Nu correlation during the pyrolysis and subsequent increase during gasification. Permeabilities increased and Dupuit-Forchheimer coefficients decreased, respectively, during and subsequently decreased and increased, respectively, during gasification. These changes were associated with the evolution of highly porous particles – also observed in the tortuosity calculations – during pyrolysis, allowing for more direct paths through the packed bed and subsequent particle break-up. Calculated effective properties showed acceptable agreement with

Chapter 10. Summary and outlook

properties available in the literature, given the differences in morphology. An in-depth understanding of the reaction mechanism was achieved and characterized.

In chapter 8, the thermal decomposition of calcium carbonate in a packed-bed configuration is introduced. It is relevant in the industrial production of calcium oxide and in CO_2 capturing processes. Its morphological and spectral radiative characterization, in the wavelength range of 0.1 to 100 µm, was carried out. In contrast to the previously investigated multi-phase media, radiation is able to partially pass through the particles in the packed bed. The spectral extinction coefficients for the void and the calcium carbonate phase were determined. The former showed to be solely dependent on morphology while the latter was strongly influenced by the intrinsic properties of calcium carbonate. Two spectral scattering coefficients for each phase were calculated, behaved complementary and showed strong dependence on the reflection behavior of the particle surface. Two spectral scattering phase functions for each phase were determined. For diffusely reflecting particles, they showed minor spectral dependence. The results were validated by comparing the calculated properties for an independently scattering particle cloud of large opaque particles with its analytical solution.

The CT-based methodology, applied and extended in this book, is widely applicable. An application in the area of environmental science and climate modeling is introduced in chapter 9. The morphological and radiative characterization (in the spectral range of 0.3 to 3 µm) of different snow layers, each composed of a different characteristic snow type, was carried out. The calculated radiative properties were incorporated in a continuum model of a 1D snow slab irradiated by diffuse or collimated radiation flux. Macroscopic optical properties of the snow layer were calculated. The results were validated by experimental transmission measurements and showed good agreement. The influence of snow morphology was analyzed by comparing the calculated macroscopic optical properties based on the exact snow morphology with the one calculated based on simplified morphologies (packed beds of spheres). Determined deviations in reflectance up to 24% implied a significant influence of snow morphology on the radiative behavior. Soot impurities in snow were included in the model and showed a reduction in the calculated reflectivity by up to 83%. In a further study, permeabilities and Dupuit-Forchheimer coefficients of these characteristic snow samples have been determined [239].

Further development of the methodology to characterize directional morphological properties in anisotropic material is proposed. The influence of particle-

particle resistance on effective heat transfer properties should be analyzed.

Figure 10.1: 3D rendering of the CT data of the alumina foam with edge length 1.2 cm (a), and 3D rendering of fibrous ceria with edge length of 1.5 mm (b).

The methodology is used in medical engineering for mechanical characterization of bone structures and can be applied straight-forwardly for stability analyses of porous media, e.g., used in solar reactors. Additionally, other physical phenomena such as sound absorption and electrical conductivity could be analyzed.

The analysis of coupled phenomena and subsequent conclusions on whether decoupling is a reasonable assumption can be investigated by the discrete-scale numerical simulations. Additionally, the influence of neglecting fluctuation terms and of closing assumptions in the volume-averaged equations can be analyzed by this method.

The continuum model developed for the solar evaporator/decomposer reactor should be extended to a three-phase model to accurately account for the evaporation process. This would allow for an in-depth understanding of the evaporation process in the porous media and for further optimization of the reactor.

The methodology can be applied to other materials of energy- and nonenergy-related areas. An alumina foam, depicted in figure 10.1.a, produced by a novel colloid-based technique [6] is considered for applications in the area of advanced insulation material and advanced materials for efficient high-temperature solid-oxide fuel cells (anode, cathode and electrolyte). Morphological characteristics, effective heat, and (multi-phase) mass transfer properties are crucial for the determination of the insulation performance as well as the performance of the fuel cell. Fibrous ceria, depicted in figure 10.1.b, is considered for application in

Chapter 10. Summary and outlook 175

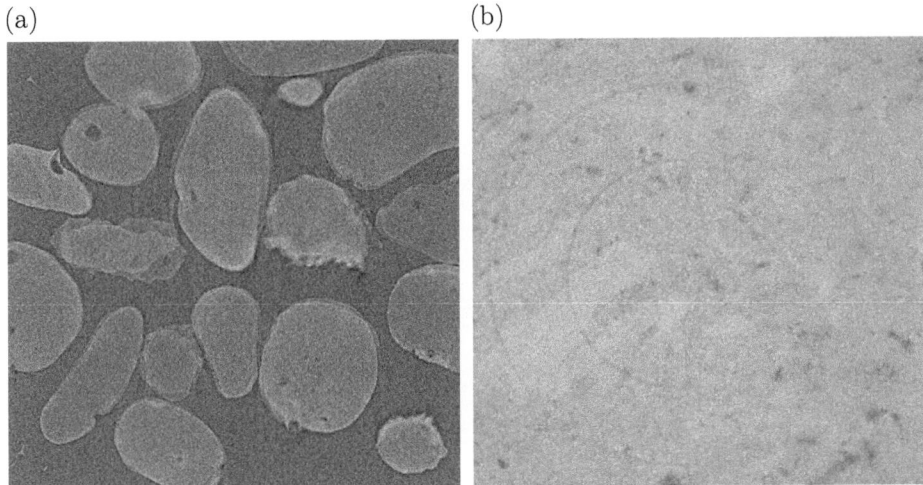

Figure 10.2: Tomographic image of a packed bed of coated polymeric particles with edge length of 1.7 mm (a), and a $La_{1.98}Sr_{0.02}CuO_4$ p-type material with edge length of 403 µm (b).

the ceria-based water-splitting process. Advantages or disadvantages of fibrous material compared to foams can be determined by morphological, heat and mass transfer characterization and, consequently, process optimization can be achieved. Packed beds of hollow particles coated with few nanometers of alumina and ferrite by atomic layer deposition [115] are depicted in figure 10.2.a. They are considered as efficient heat- and mass-transport medium for ferrite-based water-splitting cycles. Thermoelectrical processes where low-temperature solar radiation is directly converted into electrical energy [210] use p- and n-doped materials, exemplary shown for $La_{1.98}Sr_{0.02}CuO_4$ (p-type) in figure 10.2.b. Morphological characteristics and effective heat and electric transfer properties are crucial for modeling and optimization of the thermoelectric converter. Multiple applications of the CT-based method in nonenergy-related areas are possible, e.g, in medical engineering where the radiative properties of pathological and healthy tissue, exemplary shown in figure 10.3, are of interest in cancer treatment. Additional applications in medical engineering are the analysis of the mechanical stability of bone structure [144, 127], of the bone ingrowth into tissue-engineered scaffold materials [96], and of the flow and stresses within vessels [21].

The CT-based multi-scale investigations allow for an in-depth understanding

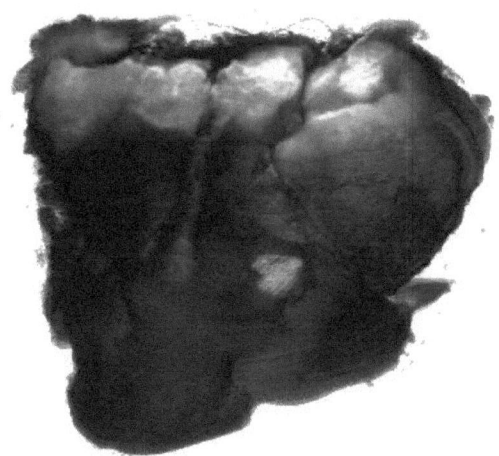

Figure 10.3: 3D rendering of CT data obtained of a mouse's skin melanoma. Edge length is approximately 6.8 mm.

of various involved processes and, additionally, help to understand the morphologies' influence on the effective heat and mass transfer properties. The preliminary studies on the tailored foam designs, described in the second part of chapter 6, are a first approach in this regard and allow for foam engineering and consequently enhanced process performance.

List of Figures

1.1 Principle of volume averaging on multi scales: a porous single particle being part of a packed bed, and the bed being part of a solar reactor. 5

2.1 Multi-component medium with component designation. 12

3.1 Scattering phase functions calculated based on the integral approach [211] or the differential approach [120, 161]. 26

3.2 Schematic of the domain, consisting of a square duct containing a sample of the multi-phase media, an inlet and outlet regions. . 29

4.1 Absorption-based CT with voxel size 3.7 µm (a) and phase-based CT with voxel size 7.4 µm (b) of animal tissue material. The edge length of the samples is 5.1 mm. 34

5.1 Equilibrium composition of sulfuric acid evaporation and decomposition at 0.1 bar (dash dotted line), 1 bar (solid line), and 10 bar (dotted line). 43

5.2 (a) Photo of the two-chamber evaporation/decomposition (left side) and SO_3 reduction reactor (right side), and (b) schematic of the evaporation chamber hosting RPC [212]. 43

5.3 Tomographic image of RPC samples: Strut of SiC RPC produced by conventional Schwartzewalder process, edge length 711 µm (a), and strut of SiSiC RPC produced by LigaFill process, edge length of 281 µm (b). 44

5.4 Tomographic image of an open-cell SiSiC RPC (a) and partially closed-cell Al_2O_3 foam (b). Edge length of the pictures are 2.3 cm. 45

5.5 Photograph of RPC sample (a), and 3D surface rendering of the RPC sample obtained by using the CT data with voxel size of 15 µm, the edge length is 3 mm (b). 46

5.6 (a) Normalized histograms of the absorption values for the HR and LR scans. The bullets indicate the corresponding threshold values of $\alpha_t/\alpha_{max} = 0.39$ and 0.23. (b) Normalized threshold absorption values for 36 subelements of three selected tomograms for voxel sizes of 30 μm and 15 μm. 46
5.7 Magnified fragment of the strut edge. 47
5.8 Opening pore size distribution of the RPC foam for the LR and HR tomography data. 48
5.9 (a) Variation in computed and measured incident radiative intensities as a function of normalized path length in the sample; (b) scattering phase functions of the RPC foam, IOTS, and of large diffuse opaque spheres as a function of the cosine of scattering angle. 49
5.10 Experimental spectroscopy setup: (1) dual Xe-Arc/Cesiwid-Glowbar lamp, (2) double monochromator, (3) and (5) collimating and focusing lens pairs, (4) sample mounted on alinear positioning stage, (6) detector, (7) optical chopper, (8) lock-in amplifier, (9) data acquisition system. 50
5.11 (a) Contour map of the normalized temperature distribution $(T - T_2)/(T_1 - T_2)$ along the axis perpendicular to the temperature boundary condition of the RPC foam (thick solid lines depict solid-fluid phase boundary) for $k_f/k_s = 1.0 \cdot 10^{-4}$; (b) the effective thermal conductivity of the RPC foam and of parallel and serial slabs at $\varepsilon = 0.91$. 51
5.12 Computed (points) and fitted (lines) Nu numbers as a function of Re and Pr numbers. 51
5.13 Normalized porosity, extinction coefficient, and effective conductivity for cubic volumes with edge lengths l. 52
5.14 Dimensionless pressure gradient as a function of Re number. . . 52
5.15 (a) Tortuosity and (b) residence time distribution for four selected Re numbers of fluid flow through the RPC foam. 53
5.16 (a) Mean residence time as a function of Re number. (b) Normalized dispersion tensor as a function of Re for the RPC foam. 53
5.17 3D model domain of foam model with foam diameter of 9 cm (a), and of reactor model with outer diameter of 25 cm (b). 54
5.18 Meassured spatial flux distribution and Gaussian fit for the measurements with a target of 0.04 m^2 and 0.55 kW total power (RMS = 5.4). 56

List of Figures

5.19 Boundary conditions for the external walls, used in the reactor model composed of solid domain (left, including vacuum domain between the window and the foam) and fluid domain (right). (1) solar irradiation inlet, (2) carrier gas inlet, (3) sulfuric acid inlet, and (4) outlet. 57

5.20 (a) Cutting plane through the middle of the foam, indicating the location of the thermocouples used for temperature measurements, and (b) recorded temperatures (left axis), acid solution inflow, and power input (right axis) of the experimental run at DLR. 58

5.21 Comparison of the experimentally obtained and the numerically determined temperatures at the positions indicated in figure 5.20. 58

5.22 Temperature distribution in the yz-plane (through the middle inlet tube) for the solid phase (left) and the fluid phase (right). 58

5.23 Temperature distribution at the outlet for the solid phase (left) and for the fluid phase (right). 59

5.24 Solid-to-fluid volumetric heat transfer along different lines, parallel to the z-axis, in the foam (a). A zoom is depicted in (b). The acid solution inlet of the middle tube is located at $y = 0.023$ m and $z = 0.015$ m. 59

5.25 Energy and chemical efficiency, and evaporation conversion as a function of total acid solution inflow (a), and inlet, interior and outlet temperature as a function of total acid solution inflow (b). 61

5.26 Energy and chemical efficiency, and evaporation conversion as a function of solar input power (a), and inlet, interior and outlet temperature as function of solar input power (b). 61

5.27 Inlet, interior and outlet temperature as a function of nominal diameter of the foam (b), and as a function of foam's porosity (b). 62

6.1 Schematic of the porous ceria-based water-splitting reactor [50]. 66

6.2 SEM pictures (HORIBA TM1000) of the graphite powder, Alfa Aeasar 40769 (a) and Alfa Aesar 10129 (b), and volume-based particle-size distributions, f, of two random samples of each graphite powder type (c). 67

6.3 CT scan (a) and 3D rendering (b) of the porous ceria foam with edge length of 376 µm. 68

6.4 Opening pore-size distribution of the three ceria samples (a) and porosity calculated for sample no. 3 in 20 subsequently growing volumes (b). The dotted horizontal line in (b) indicates $\varepsilon_{\text{num}}^2$ to which the values converge. 68

6.5 Mean intercept length for ceria sample no. 1 as a function of θ at $\varphi = 0$ and $90°$ (a), and three orthogonal planes of sample no. 1 with black bars indicating the direction of the elongated pores (b). 69

6.6 Effective normalized conductivity of the three samples as functions of k_f/k_s in the three directions and for parallel and serial slab models at $\varepsilon = 0.51$. 70

6.7 Nu number as a function of Re and Pr numbers (points) and fit (lines) for sample no. 1 in the three directions. 72

6.8 Normalized pressure drop as a function of Re number for the three directions in sample no. 1 (a), and calculated K versus F_DP for the three samples and in the three directions (b). 73

6.9 Tortuosity (a) and residence time (b) distributions for Re = 0.1 to 100 and for the three directions. 74

6.10 2D slice trough the artificial BDOTS samples with $\xi = 1$, $r_1 = 50$ μm, $\varepsilon_{\text{BDOTS}} = 0.6$ (a) and 0.8 (b). Edge length is 540 μm. ... 75

6.11 Normalized pressure gradient as a function of Re (a), and Nu numbers as a function of Re for Pr = 0.1 and 1 (b), for the artificially generated porous sample ($\xi = 1$ and $\varepsilon_{\text{BDOTS}} = 0.6$ and 0.8). 75

7.1 Equilibrium composition of stoichiometric steam gasification (a) and dry gasification (b) for 0.1 bar (dash-dotted line), 1 bar (solid line), and 10 bar (dotted line). 79

7.2 Equilibrium composition at 1 bar of steam (a,c) and dry (b,d) gasification for understoichiometric (stoichiometry factor = 0.5) (a,b) and overstoichiometric (stoichiometry factor = 2) (c,d) reactions. 79

7.3 Indirectly irradiated solar reactor used for coal gasification [166], hosting a packed bed of carbonaceous material in the reaction chamber. 82

7.4 Schematic of the tubular packed-bed reactor setup used for the gasification of carbonaceous materials. 82

List of Figures

7.5 Carbon extent X_C, starting after pyrolysis, as a function of reaction time for five different sets of process parameters, as described in table 7.2. 83

7.6 Tomograms of the reference case sample with voxel size of 3.7 µm: (a) initially at $X_C = 0$, (b) after pyrolysis at $X_C = 0.68$ (char), and (c) after gasification at $X_C = 1$ (ash). The scale bar in (c) applies to (a) and (b). In (d) a high-resolution tomogram (voxel size of 0.37 µm) of a single carbon particle at $X_C = 0.79$ is shown. 84

7.7 A 3D rendered geometry of the reference case at $X_C = 0.68$ with cube length of 1.5 cm. 85

7.8 Experimentally determined porosity for the packed bed of tire shred as a function of carbon conversion for the five experimental runs listed in table 7.2. 85

7.9 Experimentally determined specific surface area (a) and the corresponding fraction resulting from micropores (b) as a function of carbon conversion for the five experimental runs listed in table 7.2. 86

7.10 Experimentally measured and numerically calculated porosity as a function of the carbon conversion for the reference case. 87

7.11 Experimentally measured (a) and numerically calculated (b) size distribution of the particles for the reference case at various carbon conversions $X_C = 0$, 0.68, 0.79, 0.9, and 1. 87

7.12 (a) Extinction coefficient as a function of carbon conversion (markers) and its fit (solid line) given by eq. (7.10), and (b) scattering phase function for the reference case at various carbon conversions $X_C = 0$, 0.68, 0.79, 0.9, and 1 for diffusely and specularly reflecting particles. 90

7.13 Effective conductivity (normalized by solid conductivity) of the packed bed initially, after gasification (char) and after full conversion (ash) in function of the fluid to solid conductivities. Normalized effective conductivity of serial and parallel slab models are shown for the numerically determined minimum and maximum porosities of the packed bed. 91

7.14 Re-dependent Nu numbers for the packed bed at $X_C = 0$, 0.68 and 1 (initial, char and ash) and at $Pr = 0.1$, 1, and 1. The dots indicate the numerically calculated Nu numbers and the lines indicate the fits given in table 7.6. 92

7.15 Nu correlations for packed beds given by Gnielinski ($\varepsilon = 0.61$ and 0.74) [66], Wakao [225], Gunn ($\varepsilon = 0.61$ and 0.74) [71], Saidi [187], and of the present study for Pr $= 0.1$ (a), and for Pr $= 1$ (b). . . 94

7.16 Dimensionless pressure gradient, Π_{pg}, as a function of Re for the three samples of the packed bed at $X_C = 0$, 0.68 and 1 (initial, char and ash). 95

7.17 K and F_{DP} numerically determined by DPLS with those calculated by models for packed beds with $K_{inner} = 10^{-10}$ m^2 and for: $\varepsilon = 0.61$, $d = 1.14$ mm (a); $\varepsilon = 0.74$, $d = 0.97$ mm (b); $\varepsilon = 0.65$, $d = 0.42$ mm (c). F_{DP} is calculated by MacDonald (gray symbols, eq.(7.31)) and by Ward (black symbols, eq.(7.32)). 97

7.18 Residence time (a) and tortuosity (b) distribution at Re $= 1$ for the reacting packed bed at $X_C = 0$, 0.68, and 1 (initial, char, and ash). 98

8.1 Schematic of the indirectly irradiated solar calcination reactor [135]. The solar radiation is absorbed by the tubes, which emit the radiation to the calcium carbonate particles. The particles are transported through the tube and react. The products leave the reactor at the front side. 102

8.2 Sample of the packed bed of $CaCO_3$ particles: top view photograph (a), 2D tomographic image (b), and 3D surface rendering (c). 103

8.3 Normalized histogram of the sample's absorption values obtained by CT for the void phase (left peak) and for the solid phase (right peak). The bullet indicates the threshold value $\alpha/\alpha_{max} = 0.43$ used for phase identification. 103

8.4 (a) Two-point correlation function for the $CaCO_3$ packed bed. The value at $r = 0$ corresponds to the bed porosity. The dashed line indicates the asymptotic value of the function, which corresponds to ε^2. (b) Determination of the REV edge length (indicated by the vertical dashed line) by calculating the porosity of ten subvolumes with varying edge lengths l at random locations. The tolerance band for conversion and determination of the REV volume at $\varepsilon \pm 0.01$ is indicated by the two horizontal dashed lines. 104

8.5 Opening size distribution functions of the solid and fluid phases of the $CaCO_3$ packed bed ($d_h = d_{h,pore}$ for fluid and $d_h = d_{h,particle}$ for solid). 105

8.6 SEM picture of a single $CaCO_3$ particle. 105

8.7 (a) Complex refractive index of $CaCO_3$: real part obtained experimentally (solid line) [158], and imaginary part obtained by the Lorentz theory in the spectral range 0.2 to 6 µm [174] and experimentally in the remaining range (dashed line) [158]. (b) Spectral directional-hemispherical reflectivities at the specular fluid-solid interface for selected incidence directions, and spectral hemispherical reflectivity of the diffuse fluid-solid interface. . . . 106

8.8 Directional-hemispherical reflectivity as a function of incident angle at phase boundary from medium 1 to medium 2: $k_i = 0$, with $n_1 < n_2$ or $n_1 > n_2$ (a), and refraction angle as a function of incident angle from medium 1 to medium 2: $k_i = 0$, with varying $n_1 < n_2$ or $n_1 > n_2$ (b). 106

8.9 (a) Internal absorption and scattering coefficients of $CaCO_3$ particles. (b) Ratio of the scattering efficiency factor obtained for dependent scattering calculated by gas, packed-sphere, liquid, and modified-liquid models to the one obtained for independently scattering calculated by Mie theory. 107

8.10 Spectral scattering coefficients of the $CaCO_3$ packed bed for the fluid phase (a,b) and the solid phase (c,d), assuming specularly reflecting particles (a,c) and diffusely reflecting particles (b,d). . 107

8.11 Spectral extinction coefficients of the packed bed: fluid phase (a) and solid phase (b). 108

8.12 Probability density functions of the directional cosine of the incident angle at the fluid-solid interface for selected wavelengths. 108

8.13 Scattering phase functions of the $CaCO_3$ packed bed versus cosine of the scattering angle for a diffusely reflecting solid-fluid interface, at selected wavelengths $\lambda = 0.1$ µm, 1 µm, 10 µm, and 100 µm. 109

8.14 Scattering phase functions of the $CaCO_3$ packed bed versus cosine of the scattering angle for a specularly reflecting solid-fluid interface, at selected wavelengths $\lambda = 0.1$ µm, 1 µm, 10 µm, and 100 µm. 109

8.15 (a) Scattering phase function Φ_{12} and Φ_{21} for specular and diffuse fluid-solid interfaces, as a function of the scattering angle cosine, for selected refractive indices $n_2 = 1.31$, 1.64, and 1.97. (b) Normalized two-norm of the cumulative distribution functions. 110

8.16 MC and analytically calculated phase functions of a particle cloud for $f_v = 1.6 \cdot 10^{-3}$, $d = 2$ μm, $n = 1.64$, and $k = 2.6 \cdot 10^{-5}$ at $\lambda = 1$ μm, $\rho_{r,d} = 0.866$ and for specularly and diffusively reflecting particles. 111

8.17 Normalized mean intensity along three orthogonal directions as a function of sample length. 112

9.1 Rendered geometry of the snow samples investigated: ds (a), mI (b), mII (c), dh (d), and ws (e). The size of the samples (a)–(c) is 6 x 6 x 4 mm³ and the size of the samples (d)–(e) is 10.8 x 10.8 x 7.2 mm³. 116

9.2 Opening size distribution functions of the pores (a), and the particles (b) for the five characteristic snow samples. 118

9.3 Real (solid line) and imaginary (dashed line) part of the complex refractive index of ice for 0.3 μm < λ < 3 μm [228]. 119

9.4 Spectral scattering coefficients of the five characteristic snow samples for the fluid phase (a,b) and the solid phase (c,d). 119

9.5 Spectral extinction coefficients of the five characteristic snow samples: fluid phase (a) and solid phase (b). 120

9.6 Scattering phase functions of the five characteristic snow samples at $\lambda = 0.5$ μm (a) and at $\lambda = 1.5$ μm (b). 120

9.7 Schematic of the snow slab sample used for the continuum domain calculations. The inlet radiative heat flux is depicted by q''_{in}, the black body surrounding at the inlet and outlet is given by $T = 0$ K and $\varepsilon_r = 1$, respectively. The lateral walls are infinitesimally large. 120

9.8 Spectral dependence of the overall reflectance (a), transmittance (b), and absorptance (c) of a slab with $l_{slab} = 4$ cm and composed of the five characteristic snow types and irradiated by collimated radiative flux. 121

9.9 Comparison of the spectral dependence of the overall reflectance (a), and transmittance (b) of a slab with $l_{slab} = 4$ cm, composed of the five characteristic snow types and irradiated by collimated (thin solid line) and diffuse (thick solid line) radiative flux. 122

9.10 Transmittance in function of the sample's thickness for different wavelengths for two characteristic snow samples: ds snow (a) and ws snow (b). 122

9.11 Transmittance in function of the five characteristic snow samples for varying λ and for $l_{slab} = 2$ cm (a) and 8 cm (b). 122

List of Figures

9.12 Fraction of incident energy absorbed at the pore-level scale in the xy-plane ($z = 0.2$ mm) of the mII type snow for $\lambda = 1$ μm (c) and 2.5 μm (a). The corresponding snow structure is shown in (b), where black denotes void/air phase and white the ice particle. The statistically and spatial averaged absorption behavior in mII snow type for $\lambda = 1$ μm at $z = 0.2$ mm is depicted in (d). . . . 123

9.13 Fraction of incident energy absorbed at the pore-level scale in a 3 x 3 x 4 mm^3 mII type (a) and a 5.4 x 5.4 x 7.2 mm^3 ws type (b) snow sample at $\lambda = 2.5$ μm. The samples are diffusely irradiated from the front. 123

9.14 Transmittance as a function of sample thickness calculated for a ws snow sample at 825 and 925 nm (solid and dashed lines) compared to experimentally determined transmittance of a similar snow type at 830 and 927 nm (solid and dashed lines) (a); and experimentally determined transmittance of wet refrozen snow, stomped ns, and stomped ss with mean thicknesses of 4.5, 4.85 and 4.75 cm, respectively (b). 124

9.15 Comparison of R of a ds snow slab (a) and ws snow slab (b) of $l_{\text{slab}} = 4$ cm using either exact snow morphology (CT) or idealized morphology (3 types of IOSS). 126

9.16 Real (solid line) and imaginary (dashed line) part of the complex refractive index of soot for 0.3 μm $< \lambda > 3$ μm [27]. 126

9.17 Reflectance of a snow slab of $l_{\text{slab}} = 4$ cm of ds snow type (a,c) and ws snow type (b,d) containing soot impurities, either modeled as single particles homogeneously distributed within the ice (a,b) or as agglomerates composed of the same size primary particles homogeneously distributed within the ice (c,d). 128

10.1 3D rendering of the CT data of the alumina foam with edge length 1.2 cm (a), and 3D rendering of fibrous ceria with edge length of 1.5 mm (b). 136

10.2 Tomographic image of a packed bed of coated polymeric particles with edge length of 1.7 mm (a), and a La$_{1.98}$Sr$_{0.02}$CuO$_4$ p-type material with edge length of 403 μm (b). 136

10.3 3D rendering of CT data obtained of a mouse's skin melanoma. Edge length is approximately 6.8 mm. 137

List of Tables

5.1	Arithmetic mean diameter, mode, median, and hydraulic diameter for the 30 µm and 15 µm voxel size tomography data.	48
5.2	Mean values and standard deviations of the extinction coefficient along three directions.	49
6.1	Numerically and experimentally determined porosity, numerically determined specific surface, mean, mode, median and hydraulic diameter of the three porous ceria samples.	69
6.2	Numerically determined extinction coefficients of the three porous ceria samples in the three principal directions.	70
6.3	Fitting parameter a_1 of eq. 6.3.	71
6.4	Nu correlations for the three porous ceria samples in the three principal directions.	72
6.5	Calculated permeability and Dupuit-Forchheimer coefficient for the three samples in the three principal directions.	73
6.6	Mean, mode, and median tortuosity and residence time for sample no. 1 in the three directions for a sample with length 0.17mm at $Re = 1$.	74
6.7	Permeability, Dupuit-Forchheimer coefficient, and Nu correlation for the artificial porous sample with $\xi = 1$, and $\varepsilon_{\text{BDOTS}} = 0.6$ and 0.8.	75
7.1	Efficiency, specific electrical output, electrical gain factor, and the specific CO_2 emissions of the five different electricity generation routes using tire waste shreds or anthracite as feedstock and H_2O or CO_2 as gasifying agent.	81
7.2	Process parameters for the five experimental runs.	83
7.3	Resolution of tomographic scans obtained for the packed bed of carbonaceous material at different X_C.	85
7.4	Mean and median diameter of the particles in the packed bed as function of reaction extent.	88

7.5	Coefficients of the exponential fit to the scattering phase function for specularely reflecting solid-gas interface as a function of X_C.	91
7.6	Fitted Nu correlations and RMS of the fit for the packed bed at $X_C = 0$, 0.68 and 1.	92
7.7	Numerically determined Nu correlations for the three packed beds composed of IOOS with the same porosity and diameter as the reacting packed bed at $X_C = 0$, 0.68 and 1.	94
7.8	X_C-dependent permeability and Dupuit-Forchheimer coefficient of the reacting packed bed.	95
7.9	Permeability and Dupuit-Forchheimer coefficient of the three packed beds composed of IOOS with the same porosity and diameter as the reacting packed bed at $X_C = 0$, 0.68 and 1.	98
7.10	Mean, mode and median tortuosity and residence time for the packed bed at Re = 1.	98
8.1	Arithmetic mean diameter, mode, median, and hydraulic diameter calculated from the pore and particle size distributions.	105
9.1	Grain shape classification (ICSSG) [54], measured density [103], preparation method, and voxel size of the CT scans of the characteristic snow samples.	117
9.2	Calculated and measured [103] porosity and specific surface area, respectively, and REV length of the five characteristic snow samples.	118
9.3	Calculated mean, mode, median and hydraulic pore and particle diameter of the five characteristic snow samples.	118
9.4	Normalized, 2-norm ξ, and relative difference at $\lambda = 1$ µm, $r_{\text{dif},\lambda=1}$, of the calculated R based on exact morphology and idealized geometry.	126

Bibliography

[1] S. Abanades and G. Flamant. Thermochemical hydrogen production from a two-step solar-driven water-splitting cycle based on cerium oxides. *Energy*, 80:1611–1623, 2006.

[2] S. Abanades, A. Legal, A. Cordier, G. Peraudeau, G. Flamant, and A. Julbe. Investigation of the reactive cerium-based oxides for H_2 production by thermochemical two-step water-splitting. *Journal of Material Science*, 45:4163–4173, 2010.

[3] International Energy Agency. *Energy Outlook 2008*. 2008.

[4] U. S. Environmental Protection Agency. *Scrap tire technology and markets*. William Andrew Publishing/Noyes, 1993.

[5] C. Agrafiotis, M. Roeb, A. Konstandopoulos, V. Zaspalis L. Nalbandian, C. Sattler, P. Stobbe, and A. Steele. Solar water splitting for hydrogen production with monolithic reactors. *Solar Energy*, 79:409–421, 2005.

[6] J. Alder, H. Heymer, and G. Standke. Ceramic net-like structures. *Forum of Technology*, 9:19–22, 1999.

[7] J. Alder, M. Teichgraber, and G. Standke. Offenzellige Schaumkeramik mit hoher Festigkeit und Verfahren zu deren Herstelleung. *DE Patent, no. 19621638.9*, 1996.

[8] C. Argento and D. Bouvard. A ray tracing method for evaluating the radiative heat transfer in porous media. *International Journal of Heat and Mass Transfer*, 39:3175–3180, 1996.

[9] D. Bänniger, C. Bourgeois, M. Matzl, and M. Schneebeli. Reflectance modeling for real snow structures using a beam tracing model. *Sensors*, 8:10.3390/s8053482, 2008.

[10] A. Barra, G. Diepvens, J. Ellzey, and M. Henneke. Numerical study of the effects of the material properties on flame stabilization in a porous burner. *Combustion and Flame*, 134:369–379, 2003.

[11] J. Bear. *Dynamics in fluids in porous media*. Dover Publications Inc., New York, 1988.

[12] J. Berryman and S. Blair. Use of digital image analysis to estimate fluid permeability of porous material: Application of two-point correlation functions. *Journal of Applied Physics*, 60:1930–1938, 1986.

[13] A. Bhattacharya, V. Calmidi, and R. Mahajan. Thermophysical properties of high porosity metal foams. *International Journal of Heat and Mass Transfer*, 45:1017–1031, 2001.

[14] C. Bilgen and E. Bilgen. An assessment on hydrogen production using central receiver solar systems. *International Journal of Hydrogen Energy*, 9:197–204, 1984.

[15] L. Boeckx, M. Brennan, K. Verniers, and J. Vandenbroeck. A numerical scheme for investigating the influence of the three dimensional geometrical features in porous polymeric foam on its sound absorbing behaviour. *Acta Acustica united with Acustica*, 96:239–246, 2010.

[16] C. Bohren and D. Huffman. *Absorption and scattering of light by small particles*. John Wiley and Sons Inc., New York, 2004.

[17] T. Bond and W. Bergstrom. Light absorption by carbon particles: An investigative review. *Aerosol Science and Technology*, 40:27–67, 2006.

[18] K. Boomsma and D. Poulikakos. On the effective thermal conductivity of three-dimensionally structured fluid-saturated metal foam. *International Journal of Heat and Mass Transfer*, 44:827–836, 2001.

[19] K. Boomsma, D. Poulikakos, and F. Zwick. Metal foams as compact high performance heat exchangers. *Mechanics of Materials*, 35:1161–1176, 2003.

[20] M. Born and E. Wolf. *Principles of optics*. Cambridge University Press, 1999.

[21] E. Boutsianis. *Anatomically accurate hemodynamic simulations in the aorta and the coronary arteries*. PhD thesis, ETH Zurich, 2006.

[22] L. Brecher, S. Spewock, and C. Warde. The Westinghouse sulfur cycle for the thermochemical decomposition of water. *International Journal of Hydrogen Energy*, 2:7–15, 1977.

[23] L. Brecher and C. Wu. Electrolytic decomposition of water. *U.S. Patent, no. 3888750*, 1975.

[24] M. Brewster and C. Tien. Radiative transfer in packed fluidized beds – dependent versus independent scattering. *Journal of Heat Transfer*, 104:573–579, 1982.

[25] R. Buck. *Massenstrom-Instabilitäten bei volumetrischen Receiver-Reaktoren*. PhD thesis, Universität Stuttgart, 2000.

[26] C. Chan and C. Tien. Radiative transfer in packed bed spheres. *Journal of Heat Transfer*, 96:52–58, 1974.

[27] H. Chang and T. Charalampopoulos. Determination of the wavelength dependence of refractive indices of flame soot. *Mathematical and Physical Sciences*, 56:577–591, 1990.

[28] P. Charvin, S. Abanades, G. Flamant, and F. Lemort. Two-step water splitting thermochemical cycle based on iron oxide redox pair for solar hydrogen production. *Energy*, 32:1124–1133, 2004.

[29] C. Chen. Filtration of aerosols by fibrous media. *Chemical Reviews*, 55:595–623, 1955.

[30] J. Chen and S. Churchill. Radiation heat transfer in packed beds. *AIChE Journal*, 9:35–41, 1963.

[31] W. Cheuh and S. Haile. A thermochemical study on ceria: Exploiting an old material for new modes of energy conversion and CO_2 mitigation. *Philosophical Transactions of the Royal Society*, 368:3269–3294, 2010.

[32] W. Chueh and S. Haile. Ceria as thermochemical reaction medium for selectively generating syngas or methane from H_2O and CO_2. *ChemSusChem*, 2:735–739, 2009.

[33] S. Churchill and H. Chu. Correlating equations for laminar and turbulent free convection from a horizontal cylinder. *International Jornal of Heat and Mass tranfer*, 18:1049–1053, 1975.

[34] S. Churchill and H. Thelen. Eine allgemeine Korrelationsgleichung für Wärme- und Stoffübertragung bei freier Konvektion. *Chemie Ingenieur Technik*, 10:401–456, 1975.

[35] P. Coelho. Numerical simulation of the interaction between turbulence and radiation in reactive flows. *Progress in Energy and Combustion Science*, 33:311–383, 2007.

[36] R. Coquard and D. Baillis. Radiative characterization of opaque spherical particle beds: a new method of prediction. *Journal of Thermophysics and Heat Transfer*, 18:178–186, 2004.

[37] R. Coquard and D. Baillis. Radiative characteristics of beds of spheres containing an absorbing and scattering medium. *Journal of Thermophysics and Heat Transfer*, 19:226–234, 2005.

[38] R. Coquard, D. Baillis, and D. Quenard. Radiative properties of expanded polysterene foams. *Journal of Heat Transfer*, 131:012702, 2009.

[39] P. Coray, W. Lipiński, and A. Steinfeld. Experimental and numerical determination of thermal radiative properties of ZnO. *Journal of Heat Transfer*, 132:012701, 2009.

[40] F. Czechowski and H. Kidawa. Reactivity and susceptibility to porosity development of coal maceral chars on steam and carbon dioxide gasification. *Fuel Processing Technology*, 29:57–73, 1991.

[41] J. Dahl, K. Buechler, R. Finley, T. Stanislaus, A. Weimer, A. Lewandowski, C. Bingham, A. Smeets, and A. Schneider. Rapid solar-thermal dissociation of natural gas in an aerosol flow reactor. *Energy*, 29:715–725, 2004.

[42] J. Dahl, K. Buechler, A. Weimer, A. Lewandowski, and C. Bingham. Solar-thermal dissociation of methane in a fluid-wall aerosol flow reactor. *International Journal of Hydrogen Energy*, 29:725–736, 2004.

[43] W. Dalzell and A. Sarofim. Optical constants of soot and their application to heat-flux calculations. *Journal of Heat Transfer*, 91:100–104, 1969.

[44] R. Davis and H. Stone. Flow through beds of porous particles. *Chemical Engineering Science*, 48:3993–4005, 1993.

[45] C. Delisée, J. Lux, and J. Malvestio. 3D morphology and permeability of highly porous celluosic fibrous material. *Transport in Porous Media*, 91:623–636, 2009.

[46] R. Dhamrat and J. Ellzey. Numerical and experimental study of the conversion of methane to hydrogen in a porous media reactor. *Combustion and Flame*, 144:698–709, 2006.

[47] H. Douville, J.-F. Royer, and J.-F. Mahfouf. A new snow parametrization for the météo-france climate model – part 1: Validation in stand-alone experiments. *Climate Dynamics*, 12:21–35, 1995.

[48] A. Le Duigou, J.-M. Borgard, B. Larousse, D. Doizi, R. Allen, B. Ewan, G. Priestman, R. Elder, R. Devonshire, V. Ramos, G. Cerri, C. Salvini, A. Giovannelli, G. De Maria, C. Corgnale, S. Brutti, M. Roeb, A. Noglik, P.-M. Rietbrock, S. Mohr, L. de Oliveira nad N. Monnerie, M. Schmitz, C. Sattler, A. Martinez, D. de Lorenzo Manzano, J. Rojas, S. Dechelotte, and O. Baudouin. Hytec: An EC funded search for a long term massive hydrogen production route using solar and nuclear technologies. *International Journal of Hydrogen Energy*, 32:1516–1529, 2007.

[49] F. Dullien. *Porous media fluid transport and pore structure*. Academic Press, New York, 1979.

[50] C. Falter. A two-step solar thermochemical cycle based on ceria redox-reactions: Reactor design, fabrication and testing. Semester thesis, ETH Zurich, 2010.

[51] G. Faris and R. Byer. Beam-deflection optical tomography of a flame. *Journal of Energy*, 12:70–77, 1987.

[52] J. Farmer and J. Howell. Comparison of Monte Carlo strategies for radiative transfer in participating media. *Advances in Heat Transfer*, 31:333–429, 1998.

[53] T. Fend, B. Hoffschmidt, R. Pitz-Paal, O. Reutter, and P. Rietbrock. Porous materials as open volumetric solar receivers: Experimental determination of thermophysical and heat transfer properties. *Energy*, 29:823–833, 2004.

[54] C. Fierz, R. Armstrong, Y. Durand, P. Etchevers, E. Greene, D. McClung, K. Nishimura, P. Satyawali, and S. Sokratov. The international classifica-

tion of seasonal snow on the ground. IHP – VII technical documents in hydrology n 83, UNESCO-IHP, 2009.

[55] M. Flechsenhar and C. Sasse. Solar gasification of biomass using oil shale and coals as candidate materials. *Energy*, 20:803–810, 1995.

[56] J. Foster and A. Rango. Advances in modeling snowpack processes utilizing remote sensing technology. *GeoJournal*, 19:185–192, 1989.

[57] H. Friess. *A tetrahedral mesh generator based on domain indicator functions*. ETH Zurich, 2010.

[58] H. Friess, S. Haussener, J. Petrasch, and A. Steinfeld. Tetrahedral mesh generation based on space indicator functions. *Journal for Numerical Methods in Engineering*, in press, 2012.

[59] X. Fu, R. Viskanta, and J. Gore. A model for the volumetric radiation characteristics of cellular ceramics. *International Communication on Heat and Mass Transfer*, 24:1069–1082, 1997.

[60] X. Fu, R. Viskanta, and J. Gore. Prediction of effective thermal conductivity of cellular ceramics. *International Communication on Heat and Mass Transfer*, 25:151–160, 1998.

[61] P. Furler. Modeling of a high-temperature solar reactor for hydrogen production via two-step ceria cycle. Master thesis, ETH Zurich, 2010.

[62] J. Gaabab. Experimental investigation of the morphological changes in a apcked bed of tire shreds undergoing gasification. Semester thesis, ETH Zurich, 2008.

[63] M. Galvéz, A. Frei, F. Meier, and A. Steinfeld. Production of AlN by carbothermal and methanothermal reduction of Al_2O_3 in a N_2 flow using concentrated thermal radiation. *Industrial and Engineering Chemistry Research*, 48:528–533, 2009.

[64] M. Gergely, M. Schneebeli, and K. Roth. First experiments to determine snow density from diffuse near-infrared transmittance. *Cold Region Science and Technology*, 64:81–86, 2010.

[65] D. Ginosar, L. Petkovic, A. Glenn, and K. Burch. Stability of supported platinum sulfuric acid decomposition catalysts for use in thermochemical water splitting cycles. *International Journal of Hydrogen Energy*, 32:482–488, 2006.

[66] V. Gnielinski. Gleichung zur Berechnung des Wärme- und Stoffaustausches in durchströmten ruhenden Kugelschüttungen bei mittleren und grossen Pecletzahlen. *Verfahrenstechnik*, 12:363–366, 1978.

[67] N. Gokon, S. Takahashi, H. Yamamoto, and T. Kodama. Thermochemical two-step water-splitting reactor with internally circulating fluidized bed for thermal reduction of ferrite particles. *International Journal of Hydrogen Energy*, 33:2189–2199, 2008.

[68] R. Gonzalez. *Digital image processing*. Prentice-Hall, Upper Saddle River, 2002.

[69] D. Gregg, R. Taylor, J. Campbell, J. Taylor, and A. Cotton. Solar gasification of coal, activated carbon, coke and coal and biomass mixtures. *Solar Energy*, 25:353–364, 1980.

[70] D. Gregg, R. Taylor, J. Campbell, J. Taylor, and A. Cotton. Solar gasification of carbonaceous materials. *Solar Energy*, 30:513–525, 1983.

[71] D. Gunn. Transfer of heat and mass to particles in fixed and fluidize beds. *International Journal of Heat and Mass Transfer*, 21:467–476, 1978.

[72] A. Gusarov. Homogenization of radiation transfer in two-phase media with irregular phase boundaries. *Physical Review B*, 77:064202, 2008.

[73] P. Habisreuther, N. Djordjevic, and N. Zarzalis. Statistical distribution of residence time and tortuosity of flow through open-cell foams. *Chemical Engineering Science*, 64:4943–4954, 2009.

[74] A. Halmann, A. Frei, and A. Steinfeld. Carbothermal reduction of alumina: Thermochemical equilibrium calculations and experimental investigations. *Energy*, 32:2420–2427, 2007.

[75] A. Halmann and A. Steinfeld. Combined thermoneutral processes for CO_2 emission avoidance and fuel saving in the metallurgical and lime industries. *Studies in surface Science and Catalysis*, 153:481–486, 2004.

[76] J. Happel and H. Brenner. *Low Reynolds number hydrodynamics*. Martinus Nijhoff Publishers, 1988.

[77] S. Haussener. *Modeling and optimization of a high temperature solar reactor*. VDM Verlag Dr. Müller, Saarbrücken, 2008.

[78] S. Haussener, P. Coray, W. Lipiński, P. Wyss, and A. Steinfeld. Tomography-based heat and mass transfer characterization of reticulate porous ceramics for high-temperature processing. *Journal of Heat Transfer*, 132:023305, 2010.

[79] S. Haussener, M. Gergely, M. Schneebeli, and A. Steinfeld. Determination of the macroscopic optical properties of snow based on exact morphology and direct pore-level heat transfer modeling. *Journal of Geophysical Research*, submitted, 2012.

[80] S. Haussener, D. Hirsch, C. Perkins, A. Weimer, and A. Steinfeld. Modeling of a multitube high-temperature solar thermochemical reactor for hydrogen production. *Journal of Solar Energy Engineering*, 131:024503, 2009.

[81] S. Haussener, I. Jerjen, P. Wyss, and A. Steinfeld. Tomography-based determination of effective transport properties for reacting porous media. *Journal of Heat Transfer*, 134:012601, 2012.

[82] S. Haussener, W. Lipiński, J. Petrasch, P. Wyss, and A. Steinfeld. Tomographic characterization of a semitransparent-particle packed bed and determination of its thermal radiative properties. *Journal of Heat Transfer*, 131:072701, 2009.

[83] S. Haussener, W. Lipiński, P. Wyss, and A. Steinfeld. Tomography-based analysis of radiative transfer in reacting packed beds undergoing a solid-gas thermochemical transformation. *Journal of Heat Transfer*, 132:061201, 2010.

[84] S. Haussener and A. Steinfeld. Effective heat and mass transport properties of anisotropic porous ceria for solar thermochemical fuel generation. *Materials*, 5:192–209, 2012.

[85] D. Haworth. Progress in probability density function methdos for turbulent reacting flows. *Progress in Energy and Combustion Science*, 36:168–259, 2010.

[86] L. Helfen, F. Dehn, P. Mikulik, and T. Baumbach. Three-dimensional imaging of cement microstructure evolution during hydration. *Advances in Cement Research*, 17:103–111, 2005.

[87] T. Hendricks and J. Howell. Absorption/scattering coefficients and scattering phase function in reticulate porous ceramics. *Journal of Heat Transfer*, 118:79–87, 1996.

[88] G. Herman. *Fundamentals of computerized tomography*. Springer Verlag, Dordrecht, 2009.

[89] R. Hilfer. Local porosity theory for flow in porous media. *Physical Review B*, 45:7115–7121, 1992.

[90] D. Hirsch and A. Steinfeld. Solar hydrogen production by thermal decomposition of natural gas using a vortex-flow reactor. *International Journal of Hydrogen Energy*, 29:47–55, 2004.

[91] Ansys Inc. *Ansys 12.1*, 2009.

[92] Ansys Inc. *Fluent 12.0.16*, 2009.

[93] F. Incorpera and D. DeWitt. *Introduction to heat transfer*. John Wiley and Sons, New York, 1996.

[94] G. Ingel, M. Levy, and J. Gordon. Gasification of oil shales by solar energy. *Solar Energy Materials*, 24:478–489, 1991.

[95] S. Itoh. The permeability of a random array of identical rigid-spheres. *Journal of the Physical Society of Japan*, 52:2379–2388, 1983.

[96] A. jones, C. Arns, A. Sheppard, D. Hutmacher, B. Milthorpe, and M. Knackstedt. Assessment of bone ingroth into porous biomaterials using micro-ct. *Biomaterials*, 28:2491–2504, 2007.

[97] R. Jones and G. Thomas. *Materials for the hydrogen economy*. Taylor and Francis Group, Boca Raton, 2008.

[98] K. Kamiuto, M. Iwamoto, T. Nishimura, and M. Sato. Radiation-extinction coefficient of packed-sphere systems. *Journal of Quantitative Spectroscopy and Radiative Transfer*, 45:93–96, 1991.

[99] T. Kämpfer, M. Hopkins, and D. Perovich. A three-dimensional microstructure-based photon-tracking model of radiative transfer in snow. *Journal of Geophysical Research*, 112:10.1029/2006JD008239, 2007.

[100] H. Kaneko, T. Miura, H. Ishihara, S. Taku, T. Yokoyama, H. Nakajima, and Y. Tamaura. Reactive ceramics of CeO_2-MO_x (m = mn, fe, ni, cu) for

H_2 generation by two-step water splitting using concentrated solar thermal energy. *Energy*, 32:656–663, 2007.

[101] T. Kappauf and E. Fletcher. Hydrogen and sulfur from hydrogen sulfide – vi. solar thermolysis. *Energy*, 14:443–449, 1989.

[102] M. Kaviany. *Prinicples of heat and mass transfer in porous media*. Springer, New York, 1999.

[103] M. Kerbat, B. Pinzer, T. Huthwelker, H. Gäggeler, M. Ammann, and M. Schneebeli. Measuring the specific surface area of snow with x-ray tomography and gas adsorption: Comparison and implifications for surface smoothness. *Atmospheric Chemistry and Physics*, 8:1261–1275, 2008.

[104] K. Kirchart, U. Müller, H. Oertel, and J. Zierep. Axisymmetric and non-axisymmetric convection in a cylindrical container. *Acta Mechanica*, 40:181–194, 1981.

[105] M. Knackstedt, C. Arns, M. Saadatfar, T. Senden, A. Limaye, A. Sakellariou, A. Sheppard, R. Sok, W. Schorf, and H. Steiningen. Elastic and transport properties of cellular solids derived from three-dimensional tomographic images. *Proceedings of the Royal Society*, 462:2833–2862, 2006.

[106] T. Kodama, S.-I. Enomoto, T. Hatamachi, and N. Gokon. Application of an internally circulating fluidized bed for windowed solar chemical reactor with direct irradiation of reacting particles. *Journal of Solar Energy Engineering*, 130:014504, 2008.

[107] T. Kodama, Y. Kondoh, T. Tamagawa, A. Funatoh, K.-I. Shimizu, and Y. Kitayama. Fluidized bed coal gasification with CO_2 under direct irradiation with concentrated visible light. *Energy and Fuels*, 16:1264–1270, 2002.

[108] T. Kodama, H. Ohtake, K.-I. Shimizu, and Y. Kitayama. Nickel catalyst driven by direct light irradiation for solar CO_2-reforming of methane. *Energy and Fuels*, 16:1016–1023, 2002.

[109] M. Kohout, Z. Grof, and F. tStpáneki. Pore-scale modelling and tomographic visualization of drying in granular media. *Journal of Colloid and Interface Science*, 299:342–351, 2006.

[110] Ü. Köylü and G. Faeth. Radiative properties of flame-generated soot. *Journal of Heat Transfer*, 115:409–417, 1993.

[111] P. Kuhn and A. Hunt. A new solar simulator to study high temperature solidstate reactions with highly concentrated radiation. *Solar Energy Materials*, 24:742–750, 1991.

[112] S. Kuwabara. The forces experienced by randomly distributed parallel circular cylinder or spheres in a viscous flow at small Reynolds numbers. *Journal of Physical Society of Japan*, 14:527–532, 1959.

[113] C. Kyan, D. Wasan, and R. Kintner. Flow of single-phase fluids through fibrous beds. *Industrial and Engineering Chemistry Fundamentals*, 9:596–603, 1970.

[114] N. Lewis and D. Nocera. Powering the planet: chemical challenges in solar energy utilization. *Proceedings of the National Academy of Sciences of the United States of America*, 103:15729–15735, 2006.

[115] X. Liang, S. George, A. Weimer, J. Blackson, and J. Harris. Synthesis of novel porous polymer/ceramic composite material by low-temperature atomic layer deposition. *Chemistry of Materials*, 19:5388–5394, 2007.

[116] Z. Liang, WC. Chueh, K. Ganesan, SM. Haile, and W. Lipiński. Experimental determination of transmittance of porous cerium dioxide media in the spectral range 300-1100 nm. *Experimental Heat Transfer*, 24:285–299, 2011.

[117] P. Lichty, C. Perkins, B. Woodruff, C. Bingham, and A. Weimer. Rapid high temperature solar thermal biomass gasification in a prototype cavity reactor. *Journal of Solar Energy Engineering*, 132:011012, 2010.

[118] S. Lin and R. Flaherty. Design studies of the sulfur trioxide decomposition reactor for the sulfur cycle hydrogen production process. *International Journal of Hydrogen Energy*, 8:589–596, 1983.

[119] W. Lipiński, D. Keene, S. Haussener, and J. Petrasch. Continuum radiative heat transfer modeling in media consisting of optically distinct components in the limit of geometrical optics. *Journal of Quantitative Spectroscopy and Radiative Transfer*, 111:2474–2480, 2010.

[120] W. Lipiński, J. Petrasch, and S. Haussener. Application of the spatial averaging theorem to radiative heat transfer in two-phase media. *Journal of Quantitative Spectroscopy and Radiative Transfer*, 111:253–258, 2010.

[121] B. Liu, R. Hayes, Y. Yi, J. Mmbaga, M. Checkel, and M. Zheng. Three dimensional modeling of methane ignition in a reverse flow catalytic converter. *Computer and Chemical Engineering*, 31:292–306, 2007.

[122] M. Loretz, R. Coquard, D. Baillis, and D. Maire. Metallic faoms: Radiative properties/comparison between different models. *Journal of Quantitative Spectroscopy and Radiative Transfer*, 109:16–27, 2008.

[123] P. Loutzenhiser, E. Gálvez, I. Hischier, A. Graf, and A. Steinfeld. CO_2 splitting in an aerosol flow reactor via two-step Zn/ZnO solar thermochemical cycle. *Chemical Engineering Science*, 65:1855–1864, 2010.

[124] P. Loutzenhiser, O. Tuerk, and A. Steinfeld. Production of Si by vacuum carbothermal reduction of SiO_2 using concentrated solar energy. *Journal of Metals*, 9:49–54, 2010.

[125] G. Lu and D. Do. Comparison of structural models for high-ash char gasification. *Carbon*, 32:247–263, 1994.

[126] I. MacDonald, M. El-Sayed, K. Mow, and F. Dullien. Flow through porous media – ther Ergun equation revisited. *Industrial and Engineering Chemistry Fundamentals*, 18:199–208, 1979.

[127] M. Madadi, A. Jones, C. Arns, and M. Knackstedt. 3D imaging and simulation of elastic properties in porous media. *Computing in Science and Engineering*, 11:65–73, 2009.

[128] C. Manwart, S. Torquato, and R. Hilfer. Stochasitc reconstruction of sandstones. *Physical Review E*, 62:893–899, 2000.

[129] D. Marks and J. Dozier. Climate and energy exchange at the snow surface in the alpine reaction of the Sierra Nevada – 2. Snow cover energy balance. *Water Resources Research*, 28:3043–3054, 1992.

[130] D. Marks, J. Kimball, D. Tingey, and T. Link. The sensitivity of snowmelt processes to climate conditions and forest cover during rain-on-snow: A case study of the 1996 pacific northwest flood. *Hydrological Processes*, 12:1569–1587, 1998.

[131] J. Marti. Modeling of a solar evaporator reactor for the high-temperature step of a thermochemical cycle for the production of hydrogen. Semester thesis, ETH Zurich, 2010.

[132] A. Matthews. Ceramic filters for the cast metal industry. *Advances in Ceramic Materials*, 122:293–304, 1996.

[133] M. Matzl and M. Schneebeli. Measuring specific surface area of snow by near-infrared photography. *Journal of Glaciology*, 52:558–564, 2006.

[134] K. Mecke and D. Stoyan. *Morphology of condensed matter*. Springer, Berlin, 2002.

[135] A. Meier, E. Bonaldi, G. Cella, and W. Lipiński. Multitube rotary kiln for the industrial solar production of lime. *Journal of Solar Energy Engineering*, 127:386–395, 2005.

[136] A. Meier, N. Gremaud, and A. Steinfeld. Economic evaluation of the industrial solar production of lime. *Energy Conversion and Management*, 46:905–926, 2005.

[137] T. Melchior, C. Perkins, P. Lichty, A. Weimer, and A. Steinfeld. Solar-driven biochar gasification in a particle-flow reactor. *Chemical Engineering and Processing*, 48:1279–1287, 2009.

[138] T. Melchior, N. Piatkowski, and A. Steinfeld. H_2 production by steam-quenching of Zn vapor in a hot-wall aerosol flow reactor. *Chemical Engineering Science*, 64:1095–1101, 2009.

[139] Y. Michael and K. Yang. Recent developments in axial tomography for heat and fluid flow. *Experimental Thermal and Fluid Science*, 4:637–647, 1991.

[140] J. Miller, M. Allendorf, R. Diver, L. Evans, N. Siegel, and J. Stuecker. Metal oxide composites and structures for ultra-high temperature solar thermochemical cycles. *Journal of Material Science*, 43:10.1007/s1085300723547, 2008.

[141] M. Modest. *Radiative heat transfer*. Academic Press, 2003.

[142] R. Mokso, F. Marone, and M. Stampanoni. Real time tomography at the SwissLight Source. In *10th International conference on Radiation Instrumentation*, Piscataway, 2010.

[143] A. Mousa. Prediction of gas-particle heat rtansfer coefficient. *Industrial and Engineering Chemistry Process Design and Development*, 23:805–808, 1984.

[144] R. Müller and P. Rüegsegger. Analysis of mechanical properties of cancellous bone under conditions of simulated bone atrophy. *Journal of Biomechanics*, 29:1053–1060, 1996.

[145] J. Murray and E. Fletcher. Reaction of steam with cellulose in a fluidized bed using concentrated sunlight. *Energy*, 19:1083–1098, 1994.

[146] Y. Nakashima and S. Kamiya. Mathematical programs for the analysis of the three-dimensional pore connectivity and anisotropic tortuosity of porous rocks using X-ray computed tomography image data. *Journal of Nuclear Science and Technology*, 44:1233–1247, 2007.

[147] Y. Nakashima, S. Kamiya, and T. Nakano. Diffusion ellipsoids of anisotropic porous rocks calculated by X-ray computed tomography-based random walk simulations. *Water Resources Research*, 44:10.1029/2008WR006853, 2008.

[148] V. Nikulshina, C. Gebald, and A. Steinfeld. CO_2 capture from atmospheric air via consecutive CaO-carbonation and $CaCO_3$-calcination cycles in a fluidized-bed solar reactor. *Chemical Engineering Journal*, 146:244–248, 2009.

[149] A. Noglik, M. Roeb, T. Rzepczyk, J. Hinkley, C. Sattler, and R. Pitz-Paal. Solar thermochemical generation of hydrogen: Development of a receiver reactor for the decomposition of sulfuric acid. *Journal of Solar Energy Engineering*, 131:011003, 2009.

[150] A. Noglik, M. Roeb, C. Sattler, and R. Pitz-Paal. Experimental study on sulfur trioxide decomposition in a volumetric solar receiver-reactor. *International Journal of Energy Research*, 33:799–812, 2009.

[151] A. Nolin and J. Dozier. A hyperspectral method for remotely sensing the grain size of snow. *Remote Sensing and Environment*, 74:207–216, 2000.

[152] J. Norman, G. Besenbruch, L. Brown, D. O'Keefe, and C. Allen. Thermochemical water-splitting cycle, bench scale investigations, and process engineering. Final report for the period February 1977 through December 1981, 1982.

[153] A. Odgaard. Three dimensional methods for quantification of cancellous bone architecture. *Bone*, 20:315–328, 1997.

[154] D. O'Keefe, C. Allen, G. Besenbruch, L. Brown, J. Norman, R. Sharp, and K. McCorkle. Preliminary results from bench-scale testing of a sulfur-iodine thermochemical water-splitting cycle. *International Journal of Hydrogen Energy*, 5:831–892, 1982.

[155] Intergovernmental Panel on Climate Change. *IPCC Fourth Assessment Report, Climate Change 2007: The Physical Science Basis*. Cambridge University Press, 2007.

[156] T. Osinga, W. Lipiński, E. Guillot, G. Olalde, and A. Steinfeld. Experimental determination of the extinction coefficient for a packed-bed particulate medium. *Experimental Heat Transfer*, 19:69–79, 2006.

[157] Outokumpu Research Oy. *HSC chemistry 5*, 2002.

[158] E. Palik. *Handbook of optical constants in solids II*. Elsevier, New York, 1998.

[159] C. Perkins, P. Lichty, and A. Weimer. Thermal ZnO dissociation in a rapid aerosol reactor as part of a solar hydrogen production cycle. *International Journal of Hydrogen Energy*, 33:499–510, 2008.

[160] J. Petrasch. *Multi-scale analysis of reactive flow in porous media*. PhD thesis, ETH Zurich, 2007.

[161] J. Petrasch, S. Haussener, and W. Lipiński. Discrete vs continuum level simulation of radiative transfer in semitransparent two-phase media. *Journal of Quantitative Spectroscopy and Radiative Transfer*, 112:1450–1459, 2011.

[162] J. Petrasch, F. Meier, H. Friess, and A. Steinfeld. Tomography based determination of permeability, Dupuit-Forchheimer coefficient, and interfacial heat transfer coefficient in reticulate porous ceramics. *International Journal of Heat and Fluid Flow*, 29:315–326, 2008.

[163] J. Petrasch, B. Schrader, P. Wyss, and A. Steinfeld. Tomography-based determination of effective thermal conductivity of fluid-saturated reticulate porous ceramics. *Journal of Heat Transfer*, 130:032602, 2008.

[164] J. Petrasch, P. Wyss, R. Stämpfli, and A. Steinfeld. Tomography-based multiscale analyses of the 3D geometrical morphology of reticulate porous ceramics. *Journal of the American Ceramic Society*, 91:2659–2665, 2008.

[165] J. Petrasch, P. Wyss, and A. Steinfeld. Tomography-based Monte Carlo determination of radiative properties of reticulate porous ceramics. *Journal of Quantitative Spectroscopy and Radiative Transfer*, 105:180–197, 2007.

[166] N. Piatkowski and A. Steinfeld. Solar-driven coal gasification in a thermally irradiated packed-bed reactor. *Energy and Fuels*, 22:2043–2052, 2008.

[167] N. Piatkowski, C. Wieckert, and A. Steinfeld. Experimental investigation of a packed-bed reactor for the steam-gasification of carbonaceous feedstocks. *Fuel Process Technology*, 90:360–366, 2009.

[168] G. Picard, L. Arnaud, F. Domine, and M. Fily. Determining snow specific surface area from near-infrared reflectance measurements: Numerical study on the influence of grain shape. *Cold Regions Science and Technology*, 56:10–17, 2009.

[169] M. Piller, G. Schena, M. Nolich, S. Favretto, F. Radaelli, and E. Rossi. Analysis of hydraulic permeability in porous media: From high resolution X-ray tomography to direct numerical simulation. *Transport in Porous Media*, 80:57–78, 2009.

[170] B. Pinzer and M. Schneebeli. Snow metamorphism under alternating temperature gradients: Morphology and recrystalization in surface snow. *Geophysical Research Letters*, 36:10.1029/2009GL039618, 2009.

[171] S. Pope. *Turbulent flows*. Cambridge University Press, 2000.

[172] W. Press, S. Teukolsky, W. Vetterling, and B. Flannery. *Numerical recipes in C++*. Cambridge University Press, New York, 2002.

[173] Y. Qian, W. Gustafson, R. Leung, and J. Ghan. Effects of soot-induced snow albedo change on snowpack and hydrological cycle in western United States based on weather research and forecasting chemistry and regional climate simulations. *Journal of Geophysical Research*, 114:10.1020/2008/JD011039, 2009.

[174] M. Querry, G. Osborne, K. Lies, R. Jordon, and R. Coveney. Complex refractive index of limestone in the visible and near infrared. *Applied Optics*, 17:353–356, 1978.

[175] M. Quintard and S. Whitaker. One- and two-equation models for transient diffusion processes in two-phase systems. *Advances in Heat Transfer*, 23:371–464, 1993.

[176] J. Rezaiyan and N. Cheremisinoff. *Gasification technologies – A primer for engineers and scientists*. Taylor and Francis Group, Boca Raton, 2005.

[177] M. Rhode. *Introduction to particle technology*. John Wiley and Sons, Ltd, Chichester, 2008.

[178] M. Rintoul, S. Torquato, C. Yeong, D. Keane, S. Erramilli, Y. Jun, D. Dabbas, and I. Aksay. Structure and transport properties of a porous magnetic gel via X-ray microtomography. *Physical Review E*, 54:2663–2669, 1996.

[179] A. Roberts. Statistical reconstruction of three-dimensional porous media from two-dimensional images. *Physical Review E*, 56:3203–3212, 1997.

[180] C. Rodriguez-Navarro, A. Rodriguez-Navarro, K. Elert, and E. Sebastian. Role of marble microstructure in near-infrared laser-induced damage during laser cleaning. *Journal of Applied Physics*, 95:3350–3357, 2004.

[181] M. Roeb, D. Thomey, D. Graf, C. Sattler, S. Poitou, F. Pra, P. Tochon, C. Mansilla, J.-C. Robin, F. Le Naour, R. Allen, R. Elder, I. Atkin, G. Karagiannakis, C. Agrafiotis, A. Konstandopoulos, M. Musella, P. Haehner, A. Giaconia, S. Sau, P. Tarquini, S. Haussener, A. Steinfeld, S. Martinez, I. Canadas, A. Orden, M. Ferrato, J. Hinkley, E. Lahoda, and B. Wong. Hycycles – A project on nuclear and solar hydrogen production by sulphur based thermochemical cycles. *International Journal of Nuclear Hydrogen Production and Application*, 2:202–226, 2011.

[182] A. Roesch, M. Wild, R. Pinker, and A. Ohmura. Comparison of spectral surface albedo and their impact on the general circulation model simulated surface climate. *Journal of Geophysical Research*, 107:10.1029/2001JD0000809, 2002.

[183] L. Rothman. The HITRAN-2004 molecular spectroscopic database. *Journal of Quantitative Spectroscopy and Radiative Transfer*, 96:139–204, 2005.

[184] L. Ruan, H. Qi, W. An, and H. Tan. Inverse radiation problem for determination of optical constants in fly-ash particles. *International Journal of Thermophysics*, 28:1322–1341, 2007.

[185] H. Rumpf and A. Gupte. Einflüsse und Korngrössenverteilung im Widerstandsgesetz der Porengrösse. *Chemie Ingenieur Technik*, 43:367–375, 1971.

[186] J. Saggio-Woyansky, C. Scott, and W. Minnear. Processing of porous ceramics. *American Ceramic Society Bulletin*, 71:1674–1682, 1992.

[187] M. Saidi, F. Rasouli, and M. Hajaligol. Heat transfer coefficient for a packed bed of shredded material at low Peclet numbers. *Heat Transfer Engineering*, 27:41-49, 2006.

[188] A. Sakellariou, T. Senden, T. Sawkins, M. Knackstedt, M. Turner, A. Jones, M. Saadatfar, R. Roberts, A. Limaye, C. Arns, A. Sheppard, and R. Sok. An x-ray tomography facility for quantitative prediction of mechanical and transport properties in geological, biological and synthetic systems. *Proceedings of SPIE*, 5535:473–484, 2004.

[189] J. Salles, J.-F. Thovert, R. Delannay, L. Prevors, J.-L. Auriault, and P. Adler. Taylor dispersion in porous media. Determination of the dispersion tensor. *Physics of Fluids*, 5:2348–2376, 1993.

[190] R. Santoro, H. Semerjian, P. Emmerman, and R. Goulard. Oprical tomography for flow field diagnostics. *International Journal of Heat and Mass Transfer*, 24:1139–1150, 1981.

[191] L. Schunk, P. Haeberling, S. Wepf, D. Wuillemin, A. Meier, and A. Steinfeld. A receiver-reactor for the solar thermal dissociation of zinc oxide. *International Journal of Hydrogen Energy*, 130:021009, 2008.

[192] K. Schwartzwalder and A. Somers. Method of making porous ceramic articles. *U.S. Patent, no. 3090094*, 1963.

[193] C. Sergent, C. Leroux, E. Pougatch, and F. Guirado. Hemispherical-directional reflectance measurements of natural snow in the 0.9-1.45 ţ m spectral range: Comparison with adding-doubling modelling. In *Proceedings of the International Symposium on Snow and Avalanches*, Chamonix, 1997.

[194] A. Sheppard, R. Sok, and H. Averdunk. Techniques for image enhancement and segmentation of tomographic images of porous materials. *Physica A*, 339:145–151, 2004.

[195] W. Shyy, S. Thakur, H. Ouyang, J. Liu, and E. Blosch. *Computational techniques for complex transport phenomena*. Cambridge University Press, Cambridge, 1997.

[196] K. Sing, D. Everett, R. Haul, L. Moscou, R. Pierotti, J. Rouquérol, and T. Siemieniewska. Reporting physisorption data for gas-solid systems with special reference to the determination of surface area and porosity. *Pure and Applied Chemistry*, 57:603–619, 1985.

[197] T. Smith, E. Scheinder, and A. Odgaard. Star length distribution: a volume-based concept for the characterization of structural anisotropy. *Journal of Microscopy*, 191:249–257, 1998.

[198] E. Sparrow and A. Loeffler. Longitudinal laminar flow between cylinders arranged in a regular array. *American Institute of Chemical Engineers*, 5:325–330, 1959.

[199] K. Stamnes, S. Tsay, W. Wiscombe, and K. Jayaweera. Numerical stable algorithm for discrete-ordinate-method radiative transfer in multiple scattering and emitting layered media. *Applied Optics*, 27:2502–2509, 1988.

[200] M. Stampanoni, A. Groso, A. Isenegger, G. Mikuljian, Q. Chen, A. Bertrand, S. Henein, R. Betemps, U. Frommherz, P. Böhler, D. Meister, M. Lnage, and R. Abela. Trends in synchrotron-based tomographic imaging: The SLS experience. In *Proceedings of SPIE*, 2006.

[201] B. Stanmore. Review-calcination and carbonation of limestone during thermal cycling for CO_2 sequestration. *Fuel Processing Technology*, 86:1707–1743, 2005.

[202] A. Steinfeld. Solar thermochemical production of hydrogen – a review. *Solar Energy*, 78:603–615, 2005.

[203] A. Steinfeld and A. Meier. Solar fuels and materials. In C. Cleveland, editor, *Encyclopedia of Energy*, volume 5, pages 623–637. Elsevier Inc., 2004.

[204] A. Steinfeld and R. Palumbo. Solar thermochemical process technology. In R. A. Meiers, editor, *Encyclopedia of Physical Science and Technology*, volume 15, pages 237–256. Academic Press, 2001.

[205] A. Steinfeld and G. Thompson. Solar combined thermochemical processes for CO_2 mitigation in the iro, cement, and syngas industries. *Energy*, 19:1077–1081, 1994.

[206] N. Stern. *The Economics of Climate Change: the Stern Review*. Cambridge University Press, 2007.

[207] S. Stock. Recent advantages in X-ray microtomography applied to materials. *International Materials Reviews*, 53:129–182, 2008.

[208] A. Streun, A. Böge, M. Dehler, C. Gough, W. Joho, T. Korhonrn, A. Lüdeke, P. Marchand, M. Muñoz, M. Pedrozzi, L. Rivkin, T. Schlichter, V. Schlott, L. Schulz, and A. Wrulich. Comission of the Swiss Light Source. In *Proceedings of the 2001 Particle Accelerator Conference*, Piscataway, 2001.

[209] A. Studart, U. Gonzenbach, E. Tervoort, and L. Gauckler. Processing routes to macroporous ceramics: A review. *Journal of the American Ceramic Society*, 89:1771–1789, 2006.

[210] C. Suter, P. Tomeš, A. Weidenkaff, and A. Steinfeld. Heat transfer and geometrical analysis of thermoelectric converters driven by concentrated solar radiation. *Materials*, 3:10.3390/ma3042735, 2010.

[211] M. Tancrez and J. Taine. Direct identification of absorption and scattering coefficients and phase function of a porous medium by a Monte Carlo technique. *International Journal of Heat Transfer*, 47:373–383, 2004.

[212] D. Thomey, M. Roeb, P. Rietbrock, J. Säck, C. Sattler, S. Haussener, A. Steinfeld, I. Canadas, and S. Martínez. Development of a two-chamber receiver-reactor for the solar decomposition of sulphuric acid. In *Proceedings of SolarPACES 2009 Conference*, Berlin, 2009.

[213] C. Tien. Thermal radiation in packed and fluidized beds. *Journal of Heat Transfer*, 110:1230–1242, 1988.

[214] C. Tien and B. Drolen. Thermal radiation in particulate media with dependent and independent scattering. *Annual Review of Numerical Fluid Mechanics and Heat Transfer*, 1:1–32, 1987.

[215] S. Torquato and B. Lu. Chord-length distribution function for two-phase random media. *Physical Review E*, 47:2950–2954, 1993.

[216] Y. Touloukian. *Thermophysical properties of matter*. IFI/Plenum, New York, 1970.

[217] D. Trommer, F. Noembrini, M. Fasciana, D. Rodriguez, A. Morales, M. Romero, and A. Steinfeld. Hydrogen production by steam-gasification of petroleum coke using concentrated solar power – i. Thermodynamic and

kinetic analyses. *International Journal of Hydrogen Energy*, 30:605–618, 2005.

[218] H. Uchiyama, M. Nakajima, and S. Yata. Measurement of flame temperature distribution by IR emission computed tomography. *Applied Optics*, 24:4111–4116, 1985.

[219] L. van Brakel. Mercury porosimeter: State of the art. *Powder Technology*, 29:1–12, 1981.

[220] G. Vignoles, O. Coindreau, A. Ahmadi, and D. Bernard. Assessment of geometrical and transport properties of a fibrous C/C composite preform as digitalized by X-ray computerized microtomography: Part ii. heat and gas transport properties. *Journal of Material Research*, 20:10.1557/JMR.2005.0311, 2007.

[221] X. Vitart, A. Le Duigou, and P. Carles. Hydrogen production using the sulfur-iodine cycle coupled to a VHTR: An overview. *Energy Conversion and Management*, 47:2740–2747, 2006.

[222] B. von Setten, J. Bremmer, S. Jelles, M. Makkee, and J. Moulijn. Ceramic foams as a potential molten salt oxidation catalyst support in the removal of soot from diesel exhaust gas. *Catalysts Today*, 53:613–621, 1999.

[223] P. von Zedtwitz and A. Steinfeld. The solar thermal gasification of coal – energy conversion efficiency and CO_2 mittigation potential. *Energy*, 28:441–456, 2003.

[224] R. vonZedtwitz, W. Lipiński, and A. Steinfeld. Numerical and experimental studyof gas-particle radiative heat exchange in a fluidized-bed reactor for steam-gasificaiton of caol. *Chemical Engineering Science*, 62:599–607, 2007.

[225] N. Wakao. *Heat and mass transfer in packed beds*. Gordon and Breach Science Publishers, 1982.

[226] J. Wards. Turbulent flow in porous media. *Journal of the American Society of Civil Engineers*, 90:1–12, 1964.

[227] S. Warren. Optical properties of snow. *Review of Geophysics and Space Physics*, 20:67–89, 1982.

[228] S. Warren and E. Brandt. Optical constants of ice from the ultraviolet to the microwave: A revised compilation. *Journal of Geophysical Research*, 113:10.1029/2007JD009744, 2008.

[229] S. Warren and W. Wiscombe. A model for the spectral albedo of snow – i. Snow containing atmospheric aerosols. *Journal of Atmospheric Sciences*, 37:2734–2745, 1980.

[230] A. Weidenkaff, P. Nüesch, A. Wokaun, and A. Reller. Mechanics studies of the water-splitting reaction for producing solar hydrogen. *Solid State Ionics*, 101:915–922, 1997.

[231] Z. Weszka. A survey of threshold selection techniques. *Computer Graphics and Image Processing*, 7:259–265, 1978.

[232] S. Whitaker. The Forchheimer equation: A theoretical development. *Transport in Porous Media*, 25:27–61, 1996.

[233] S. Whitaker. *The method of volume averaging, theory and applications of transport in porous media*. Kluwer Academic Publisher, Dordrecht, 1999.

[234] W. Wiscombe and S. Warren. A model for the spectral albedo of snow – i. Pure snow. *Journal of Atmospheric Sciences*, 37:2712–2733, 1980.

[235] W. Yang. *Handbook of fluidization and fluid-particle system*. Marcel Dekker, New York, 2003.

[236] Y. Yang, J. Howell, and D. Klein. Radiative heat transfer through a randomly packed bed of spheres by the Monte Carlo method. *Journal of Heat Transfer*, 105:325–332, 1983.

[237] L. Younis and R. Viskanta. Experimental investigation of the columetric heat transfer coefficient between stream of air and ceramic foam. *International Journal of Heat and Mass Transfer*, 36:1425–1434, 1993.

[238] B. Zeghondy, E. Iacona, and J. Taine. Determination of the anisotropic radiative properties of a porous material by radiative distribution function identification (rdfi). *International Journal of Heat and Mass Transfer*, 49:2810–2819, 2006.

[239] E. Zermatten, S. Haussener, M. Schneebeli, and A. Steinfeld. Tomography-based determination of permeability and dupuit-forchheimer coefficient of characteristic snow samples. *Journal of Glaciology*, 57:811–816, 2011.

[240] A. Z'Graggen. *Solar gasification of carbonaceous materials – Reactor design, modeling and experimentation.* PhD thesis, ETH Zurich, 2008.

[241] A. Z'Graggen, P. Haueter, G. Maag, A. Vidal, M. Romero, and A. Steinfeld. Hydrogen production by steam-gasification of petroleum coke using concentrated solar power – iii. Reactor experimentation with slurry-feeding. *International Journal of Hydrogen Energy*, 32:992–996, 2007.

[242] A. Z'Graggen, P. Haueter, D. Trommer, M. Romero, J. de Jesus, and A. Steinfeld. Hydrogen production by steam-gasification of petroleum coke using concentrated solar power – ii. Reactor design, testing, and modeling. *International Journal of Hydrogen Energy*, 31:797–811, 2006.

i want morebooks!

Buy your books fast and straightforward online - at one of world's fastest growing online book stores! Environmentally sound due to Print-on-Demand technologies.

Buy your books online at
www.get-morebooks.com

Kaufen Sie Ihre Bücher schnell und unkompliziert online – auf einer der am schnellsten wachsenden Buchhandelsplattformen weltweit! Dank Print-On-Demand umwelt- und ressourcenschonend produziert.

Bücher schneller online kaufen
www.morebooks.de

 VDM Verlagsservicegesellschaft mbH
Heinrich-Böcking-Str. 6-8
D - 66121 Saarbrücken

Telefon: +49 681 3720 174
Telefax: +49 681 3720 1749

info@vdm-vsg.de
www.vdm-vsg.de

Printed by Books on Demand GmbH, Norderstedt / Germany